AUSTRALIAN CRUSTACEANS IN COLOUR

Frontispiece: A burrowing shrimp, *Laomedia healyi,* from mangrove swamps, x8.

AUSTRALIAN CRUSTACEANS
IN COLOUR

Anthony Healy

John Yaldwyn

A. H. & A. W. REED

SYDNEY • MELBOURNE • WELLINGTON • AUCKLAND

This book is dedicated to those whose work on Crustacea at the Australian Museum, or in association with the Museum, has made such general accounts as this possible:

William Aitcheson Haswell, 1854–1925
Thomas Whitelegge, 1850–1927
Allan Riverstone McCulloch, 1885–1925
Charles Melbourne Ward, 1903–1966
Francis Alexander McNeill, 1896–1969

First published 1970

A. H. & A. W. REED

51 Whiting Street, Artarmon, Sydney
357 Little Collins Street, Melbourne
182 Wakefield Street, Wellington
29 Dacre Street, Auckland

© 1970 ANTHONY HEALY–JOHN YALDWYN

ISBN 0 589 07073 8

Type set by Craftsmen Type-Setters Pty Ltd, Sydney, Australia
Printed and bound by Kyodo Printing Company Limited, Tokyo, Japan

Introduction

MANY THOUSANDS of different crustaceans occur on the shores and in the seas around Australia, in the freshwaters of the continent itself and to a certain extent on the dry land away from water. The variation in size, colour and form shown by these Australian crustaceans is truly fantastic. In size and form they range from water fleas and copepods, less than a sixteenth of an inch in length, through a variously-sized series of shrimp-like forms, barnacles and crayfish, up to the giant Tasmanian deepwater crab, *Pseudocarcinus gigas,* reaching at least 30 pounds in weight and 16 inches across the carapace. In colour they range from uniform scarlet in some deepwater prawns, through green in some freshwater crayfish, to sky blue in soldier crabs. Other representatives show startling and contrasting colour patterns, as for example the symmetrical markings of the ornate marine crayfish (page 7) and the body banding of the zebra shrimp (plate 20).

Crustaceans, like insects, spiders, centipedes and related forms, belong to a major division of the animal kingdom, the Arthropoda or 'jointed-feet' invertebrates. Arthropods in general owe their great success as an animal group in large part to the development of their outer body layer, or cuticle, into protective armour. This armour, the 'crust' part of the word crustacean, is non-living and is secreted by the underlying tissue. In whole or in part it forms protective coverings, biting jaws, eye lenses, walking legs and many other external structures.

There are several layers in the arthropod cuticle. Depending on the characteristic habits and habitat of the particular representatives examined, these layers contain differing amounts of a water-proofing wax, a flexible horny substance called chitin and often rigid lime or calcium carbonate. This rigid armour, or exoskeleton, acts as a supporting framework for the tissues within and for the attachment of muscles. Between protective areas of inflexible hardening, the cuticle remains as a flexible membrane or joint, thus providing for movement. In order to grow, an arthropod must periodically shed, or moult, its outermost layers of cuticle. Underneath these the animal already possesses a newly-formed and larger cuticular sheath, which remains elastic or 'soft-shelled' for a short period after each moult to allow for a small increase in size.

It is rather difficult to define the Crustacea exclusively, as a group, so that the casual observer can immediately distinguish them from other arthropods. In general they can be recognised by their possession of gills and by the fact that, though some subgroups have terrestrial representatives, the majority are aquatic. Freshwater forms are common, but the group is predominantly marine and is so abundant in the oceans that crustaceans have sometimes been referred to as water-breathing 'insects' of the sea. A more explicit definition, which includes parasitic and modified forms as well as certain primitive and simple forms, owes much to the words of a nineteenth century English clergyman, the Rev. T. R. R. Stebbing, and can be stated as follows: Crustacea are arthropods with a segmented body and limbs at some stage in their life history; they either have gills or else breathe in water through their skins; they have two pairs of feelers, or antennae; they have no true neck; they never have wings; they hatch as free-moving larvae, and they have a chitinous cuticle which may be flexible or may, by the presence of calcium carbonate, become as hard as bone or brick.

The Crustacea can be divided into a number of major subgroups, each with its own distinguishing set of characters. Thus the diminutive, and mainly freshwater, fairy shrimps, shield shrimps and water fleas are grouped together as the Branchiopoda at one end of the scale, while the varied, and mainly marine, shrimps, crayfish, hermit crabs and true crabs are grouped together as the Decapoda at the other. The former have a series of foliaceous, or leaf-like, feet acting as gills for a common feature, while the latter all possess five pairs of thoracic legs and a well-developed carapace or shield covering the fused head and thorax. There are between eight and about thirty-six of these major crustacean subgroups recognised, depending on the level of classification used and the particular system of classification followed. Not all of these groups occur in Australia and not all those that do are mentioned in the following pages. No attempt has been made to be inclusive; the obvious, common or colourful groups have been selected for illustration and discussion, while other smaller groups, not readily available to the authors, have been omitted.

The Australian crustacean fauna has many unusual and unique features. For example, there is the mass hatching, swarming and subsequent death of shield shrimps and shelled fairy shrimps in the arid inland of the continent after occasional rain; there are 'living fossil', anaspidacean 'mountain shrimps' of great evolutionary interest in south-eastern Australia; there are terrestrial amphipods, abundant and important as leafmould animals in

Australian tropical and temperate rain forests, but virtually unknown as a group in most other areas of the world; there are numerous species of penaeid prawns supporting an extensive prawn fishery; there is a fantastic array of freshwater crayfish, including both the largest and smallest freshwater crayfish in the world; there is a giant deepwater crab which must be considered one of the largest known crustaceans, and there are gregarious soldier crabs forming characteristic and colourful 'armies' on many Australian sandy beaches.

In the following pages we deal with a series of different crustacean groups and we attempt, using colour photographs, an explanatory text and black and white illustrations, to give some idea of the great variety and spectacular appearance of this diverse group of invertebrates in Australia.

Acknowledgments

We wish to thank the following for help during the preparation of this book: Brian Bertram for the skill and patient care applied to the line drawings; J. Beeman, G. Biddle, Julie Booth, M. Cameron, H. Cogger, N. Coleman, E. Slater, Mrs J. Finch, Dr D. D. Francois, K. Gillett, D. Henderson, Janet Holloway, W. Ivantsoff, Dr Hilary Jolly, C. J. Lawler, F. G. Myers, R. Oldfield, Dr Patricia Ralph, N. Ruello, I. Smith, Mrs C. Wright (collection and photography of crustaceans); and the late Frank McNeill; Dr W. D. Williams (shield shrimps); B. V. Timms (water fleas and copepods); Miss Elizabeth Pope (barnacles); Dr D. E. Hurley (isopods and amphipods); Dr J. L. Barnard (amphipods); Dr J. C. McCain (skeleton shrimps); Dr and Mrs A. H. Banner (snapping shrimps); E. F. Riek (freshwater crayfish); Miss Janet Haig (lobster krill and half crabs); Dr D. J. G. Griffin (crabs); Dr J. S. Garth (crabs); Dr J. A. Bishop (freshwater crabs) and Dr L. B. Holthuis (decapods).

Our special thanks go to Dorothy Healy for her patience and help during the photography of assorted living and preserved crustaceans over many years and to Barbara Yaldwyn for her enthusiastic encouragement and detailed editorial help during the entire production of this work.

Books on Australian Crustacea

There is no single book dealing with the Crustacea of Australia as a whole, the nearest approach being H. M. Hale's *The Crustaceans of South Australia*, Government Printer, Adelaide, part I 1927, part II 1929. Other books with useful sections on Australian Crustacea include the companion volumes in the present series, *Australian Seashores in Colour* by Keith Gillett and John Yaldwyn and *The Australian Great Barrier Reef in Colour* by Keith Gillett, as well as the following: W. D. Williams, *Australian Freshwater Life*, Sun Books, 1968; W. J. Dakin, *Australian Seashores*, Angus & Robertson, revised edition 1966; K. Gillett and F. McNeill, *The Great Barrier Reef and Adjacent Isles*, Coral Press, third edition 1967, and Isobel Bennett, *The Fringe of the Sea*, Rigby, 1966. The general work by S. W. Tinker, *Pacific Crustacea*, Charles E. Tuttle, 1965, illustrates many characteristic tropical species found on the Great Barrier Reef.

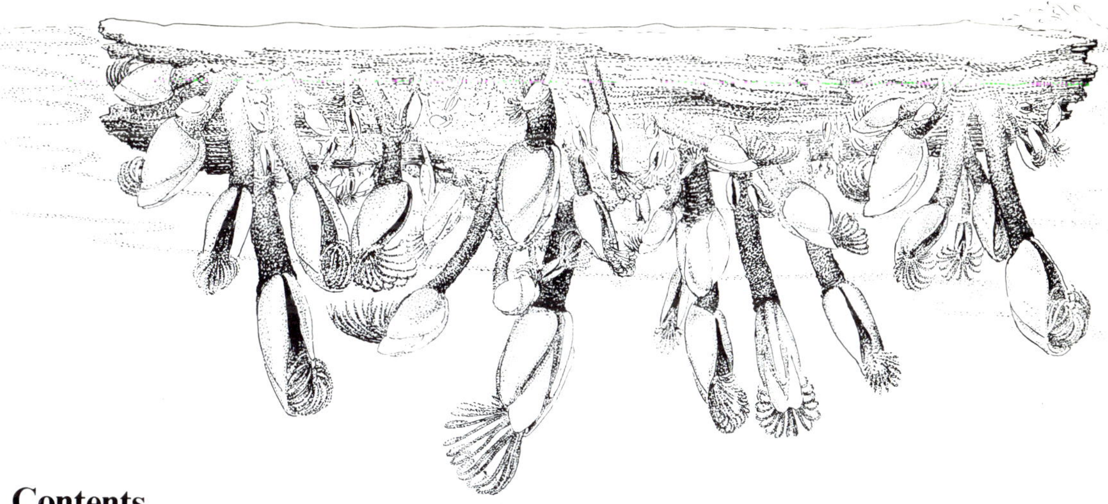

Contents

BRANCHIOPODA—Shield Shrimps and Water Fleas

THE BRANCHIOPOD or 'gill-footed' crustaceans are usually regarded as the most primitive of the lower Crustacea. 'Primitive' here implies simplicity of organisation and a general lack of the strict fixity of structure so characteristic of other groups. The term 'lower Crustacea' is used for the less specialised and relatively simple branchiopods, copepods, and cirripedes to distinguish them from the 'higher Crustacea', or Malacostraca, with their fixed pattern of body segmentation and their highly complex form.

There are four main subgroups of branchiopods: the Anostraca or 'shell-less' fairy shrimps, the Notostraca or 'shell-backed' shield shrimps, the Conchostraca or 'mollusc-shelled' fairy shrimps and the Cladocera or water fleas. The first three groups, with the exception of a few fairy shrimps, are all restricted freshwater forms. The majority of the Cladocera are also freshwater forms but several species are abundant in the sea. Almost the only feature common to all four of these groups is their possession of leaf-like, lobed feet, each bearing a gill plate functioning as a respiratory organ.

Fairy shrimps are elongate branchiopods without a shell or carapace but with a distinct head, thorax and abdomen. They have prominent stalked eyes, reach an inch or more in length and always swim upside down, that is, with their under surface and feet uppermost. About thirty species have been recorded from Australia, all but the cosmopolitan brine shrimp, *Artemia salina,* being restricted to this area.

Notostracan shield shrimps have a shield-shaped carapace over the head and part of the thorax. They have prominent, button-like eyes on the upper surface of the carapace (pl. 3), a large number of body segments and numerous, closely-spaced gill-feet. Both Australian species are large and grow to about three inches in length. The olive green *Triops australiensis* is found in temporary rain-pools and water-filled claypans in the hotter and more arid parts of Australia, while the bright green *Lepidurus viridis* occurs in ponds and drainage ditches in cooler and wetter areas such as south-western and south-eastern Australia including Tasmania.

Shelled fairy shrimps or conchostracans have a bivalved carapace enclosing the whole animal. They are usually small in size, though *Limnadopsis birchii* (fig. 1) from inland Australia reaches an inch or more in length. There are more than twenty species known from lakes, dams and water-holes throughout the continent.

Cladoceran water fleas are small and relatively

Fig. 1. Shelled fairy shrimps, *Limnadopsis birchii,* from an inland claypan. Right specimen has one shell of pair turned back to reveal body and limbs, x1.

Plate 3. Shield shrimps, *Triops australiensis,* in water-filled, desert claypan, x2. *Photo H. Cogger.*

transparent animals common in inland ponds or slow-flowing waters. The wide-ranging Australian *Daphnia carinata* (pl. 4 and fig. 2) is typical of the group. It is about one eighth of an inch in length and normally colourless, though the living animal in plate 4 has been photographed with coloured light to bring out details of internal structure. A thin-walled carapace covers the trunk and abdomen, leaving the head, bearing a pair of antennae used as swimming organs, free. Over seventy species of water fleas occur in Australia but identification is difficult as seasonal variation in shape is common in this group.

Most branchiopods can produce special drought-resistant 'resting eggs' capable of withstanding unusual heat, cold, or prolonged desiccation. For inhabitants of temporary pools and bodies of water which dry up completely, these constitute the sole means of tiding the species over from one wet season to the next. Such sexually-produced, resting eggs, as distinct from normal 'favourable-season' eggs (often parthenogenetic or unfertilised), are effective dispersal mechanisms for these forms. Minute dried eggs may be blown over great distances in dust or possibly transported by birds and insects. Resting eggs of some species hatch without drying, but in most, drying is a prerequisite for hatching.

In favourable conditions development takes place very rapidly, as Professor Baldwin Spencer observed during the Horn Expedition to central Australia in 1894. The shield shrimp *Triops australiensis* was found in hundreds in small claypans and pools in creek beds, in fact anywhere where there was water clouded with mud particles. Ever since then, *Triops* has been the characteristic crustacean seen by travellers in central Australia; as it flounders in the mud of a drying claypan its fancied resem-

blance to the completely unrelated and extinct fossil trilobites has been remarked upon by many observers. Spencer recorded the rapid growth of *Triops*. Not more than two weeks after rain fell, innumerable shield shrimps and other branchiopods were swimming in the pools. As none had been found in the area prior to the rain, all must have developed in that short time from resting eggs.

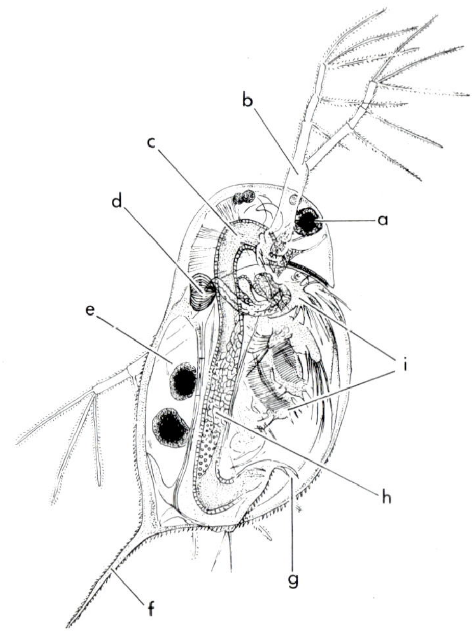

Fig. 2. Water flea, *Daphnia carinata*, drawn as key to plate 4 *(opposite)*. *a*, eye; *b*, right antenna (left folded back behind body); *c*, gut; *d*, heart; *e*, brood chamber and eggs; *f*, posterior spine; *g*, caudal, or 'tail', claws; *h*, ovary; *i*, legs.

Plate 4. Water flea, *Daphnia carinata*, alive, artificially coloured, x40. *Photomicrograph R. Oldfield.*

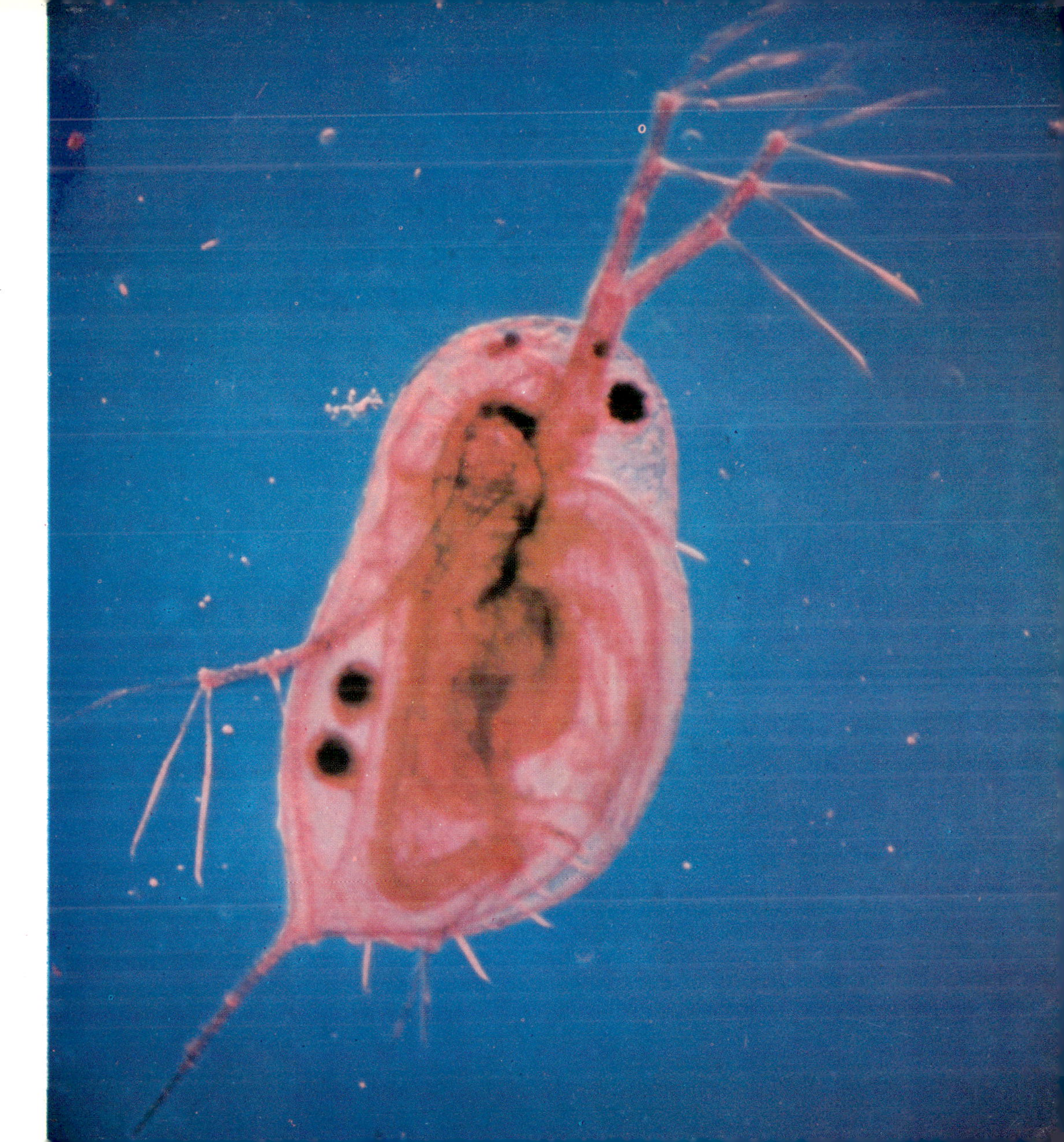

COPEPODA—'Oar-footed' Crustaceans

PROBABLY THE MOST abundant of all crustacean groups are the copepods or 'oar-footed' crustaceans. The free-living forms of this group are minute, and swarm in freshwaters and in the sea, while the numerous parasitic members are widespread and reach in some cases a foot or more in length. Copepods in general are elongated and somewhat pear-shaped in appearance, with a well-defined head and a narrow abdomen ending in a caudal fork. The first pair of antennae are enlarged and used for locomotion. They are usually held at right angles to the body and give a very characteristic and diagnostic, T-shaped appearance to many of the free-living forms (pl. 5). The eye, if present, is single and unpaired; in fact one of the earliest known copepods was given the generic name *Cyclops*. Many parasitic forms are degenerate and their structure can become highly modified. In virtually all copepods, both free-living and parasitic, no matter how modified, the eggs are carried in one or a pair of egg-sacs attached to the abdomen of the female (fig. 3).

In Australia, freshwater copepods are common, occurring at the edge of running water and in the open waters of lakes, pools and dams as well as on the bottom and among aquatic vegetation. *Boeckella* is a characteristic genus south of the tropics and the species photographed alive with coloured light in plate 5 is the widespread Australian and New Zealand *Boeckella triarticulata*.

Planktonic, or free-drifting, marine copepods are of fundamental importance in the general economy of the sea. In their myriads they feed on diatoms and other single-celled marine plants and in their turn form food for the young stages of many fishes. In the food chain of the tiger flathead, for example,

Fig. 3. Freshwater copepod, *Boeckella,* with egg-sac attached to abdomen, x35.

several copepods, including *Calanoides brevicornis,* are vital links between diatoms and the small fish which form the principal food of the flathead itself.

   Plate 5. Freshwater copepod, *Boeckella triarticulata,* alive, artificially coloured, x50. *Photomicrograph R. Oldfield.* 

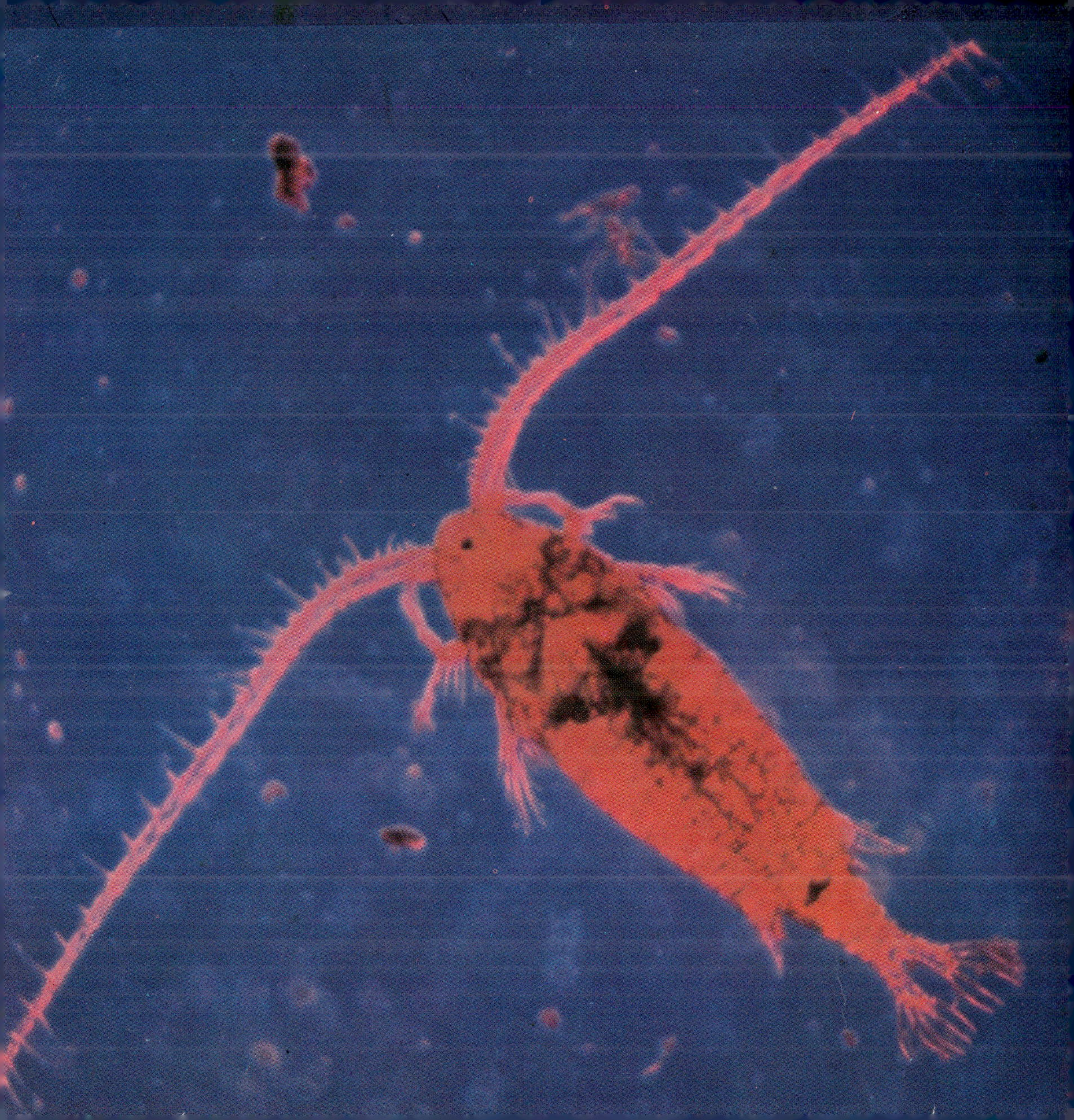

CIRRIPEDIA—Goose and Acorn Barnacles

THE MOST OBVIOUS and abundant crustaceans seen on Australian seashores are the barnacles or cirripedes. These distinctive-looking animals, with their thick calcareous plates and their invariable attachment to some fixed or free-moving object, were long regarded as molluscan shellfish. Their structure and their larval life history, however, reveal their true relationship to crustacean arthropods. Their feet, technically termed cirri (fig. 6), are jointed and their early free-swimming larval stage, the nauplius, has appendages and a single median eye, just as it does in many other crustaceans. The group is exclusively marine, though some species can survive in brackish water.

There are three main types of barnacles: goose or stalked barnacles, acorn or sessile barnacles and parasitic barnacles. The former are usually oceanic and examples can be found washed ashore, attached by a flexible stalk, or peduncle, to a log or some other floating object (pl. 6). The peduncle in some species, such as the virtually cosmopolitan *Lepas anatifera* (fig. 4), may reach a length of eighteen inches. Other, unusually short-stalked, forms live attached to intertidal and immediately subtidal rocks. The name 'goose' barnacle is derived from a mediaeval myth which regarded stalked barnacles attached to floating tree trunks as the fruit of the trees themselves and assumed that, had the tree remained growing, each fruit would have developed into a sea bird called the barnacle goose.

Acorn barnacles, in contrast, are stalkless and are attached directly by the shell to intertidal rocks or mangrove roots, though some species live attached to whales, turtles or even other crustaceans (pl. 41). The external wall, or shell, of a sessile barnacle is made up of a base (fig. 6), then a circular wall of four, six or eight plates and finally two pairs of plates closing the aperture itself. An exceptional species is the surf barnacle, *Catophragmus polymerus* (pl. 8), which has numerous smaller, overlapping accessory plates in addition to eight main wall plates. Within the shell the body of the barnacle lies, so to speak, on its back with the mouth and the six pairs of feet directed upwards towards the slit-like opening between the shell valves closing the aperture. The adductor and the retractor muscles control the opening and closing of the valves and the projection of the feet during feeding. The great nineteenth century populariser of biology, T. H. Huxley, described sessile barnacles in an often-quoted phrase as lying on their backs, holding on by

Fig. 4. Goose barnacles, *Lepas anatifera,* attached to glass fishing float.

Plate 6. Oceanic goose barnacles, *Lepas pectinata,* on floating *Janthina* shell, x8.

their heads and kicking their food into their mouths with their heels.

The parasitic barnacles vary considerably in their organisation and in their degree of degeneration. Some bore into corals and large shells, leaving a characteristic keyhole-shaped channel to the outside, while others are reduced to mere limbless sacs attached beneath the abdomen of crabs and certain other crustacean groups. *Sacculina* is perhaps the best known of these degenerate forms. The body sac may be so large that an affected crab appears superficially to be carrying a batch of eggs beneath the abdomen. The parasite has no alimentary system; nourishment is obtained through a threadlike absorptive root system which penetrates the body of the host in all directions. The effect of this internal disorganisation on a male crab is profound, resulting in a form of desexing called parasitic castration. A male crab in this condition takes on some or all of the secondary female characters of its species including the broader female abdomen and the differently-shaped female abdominal swimmerets. *Sacculina*-affected blue swimming crabs are sometimes seen in commercial catches. There is no doubt that these degenerate, limbless parasites are barnacles as their eggs hatch into characteristic barnacle nauplii, which then develop as typical free-swimming barnacle larvae until they settle on a suitable crustacean. At that stage the larva works its way into the body of its host, undergoes fundamental changes in form and finally positions itself beneath the abdomen ready to protrude through the cuticle as an external sac.

Acorn and goose barnacles feed on floating particles of food by sweeping the water rhythmically with their six pairs of double-branched feet. Each branch of this 'plankton net', a total of twenty-four, is multiarticulate and minutely haired. When submerged, the shell valves open, the feet are extended

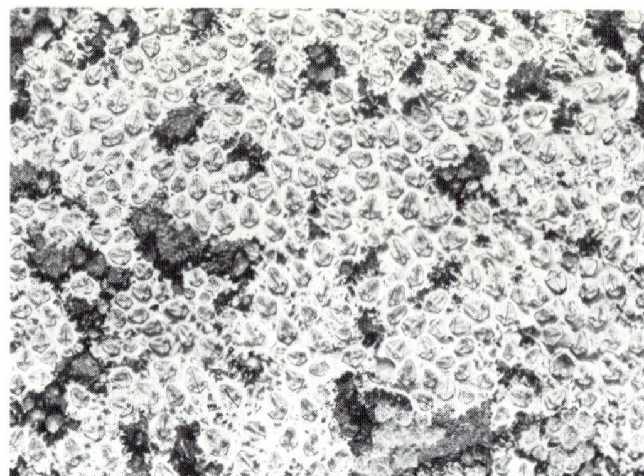

Fig. 5. Dense covering of honeycomb barnacles, *Chamaesipho columna,* on intertidal rock surface, x½.

fully and then retracted in a single coordinated sweeping stroke (fig. 8). This feeding movement is repeated rhythmically and continuously by a healthy barnacle whenever it is undisturbed and under water. Plate 6 shows a group of goose barnacles with their respective sets of feet in various stages of extension and retraction during feeding.

In Australian waters one of the common oceanic goose barnacles is the long-stalked *Lepas anatifera* (fig. 4) and logs or other floating debris may come ashore literally covered with this passively-drifting form or the closely-related, short-stalked *Lepas anserifera.* Another related form, shown here attached to the planktonic violet snail, *Janthina* (pl. 6), is *Lepas pectinata* with its characteristically serrate keel plate, or carina, and its ridged shell plates. In contrast to these goose barnacles attached to acquired floats, there is a little, planktonic barnacle, *Lepas fasicularis,* attached in small groups to shared self-secreted floats. This barnacle is blue in colour and is occasionally cast ashore on Australian

   Plate 7. Intertidal rock platform whitened with honeycomb barnacles, *Chamaesipho columna.*

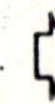

beaches with other free-floating planktonic animals, likewise blue in colour. These 'blue-floaters' include colonial coelenterates such as Portuguese men-of-war *(Physalia physalis)*, by-the-wind sailors *(Velella lata)* and their sail-less circular relative *Porpita pacifica*, as well as molluscan shellfish such as *Janthina* snails and a little, frill-bearing nudibranch called *Glaucus lineatus*.

Recent observations have revealed the remarkable growth rate of some goose barnacles under suitable conditions. A newly-painted buoy put down off the New South Wales coast in December, 1960, was retrieved seventeen days later. In that time, free-swimming larval stages of *Lepas anatifera* had settled and grown to just under an inch in shell length. These barnacles were then mature adults ready to release eggs from their ovaries. Another species of goose barnacle, *Conchoderma virgata*, which also settled on the buoy, had larvae hatching

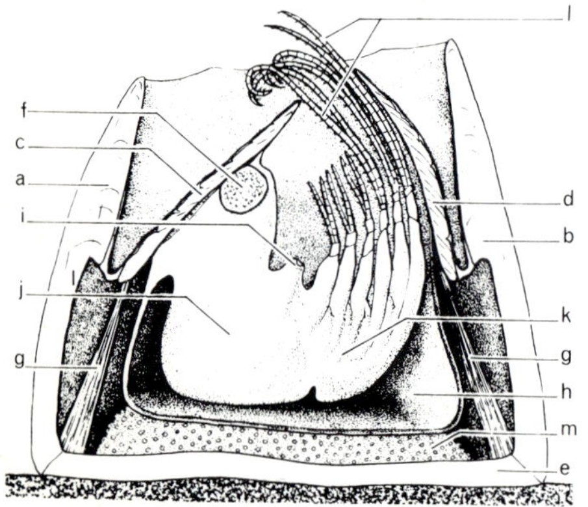

Fig. 6. An acorn barnacle cut through to show internal structure. *a-e*, shell plates; *a*, rostrum; *b*, carina; *c*, scutum; *d*, tergum; *e*, shell base; *f*, adductor muscle (cut section); *g*, retractor muscle; *h*, mantle cavity; *i*, mouth; *j*, region of stomach; *k*, thorax; *l*, cirri, or feet; *m*, ovary.

and second batches of eggs ready to be released within this seventeen-day period.

On Australian rocky shores acorn barnacles are obvious and important zonation-marking animals. That is to say they tend to occur at very definite and often clearly marked intertidal levels depending on the interaction of a whole series of factors, such as desiccation, exposure to air, exposure to wave action, interrelations with other animals and plants, and the actual slope of the intertidal rock surface itself. Thus in the eastern Australian, warm temperate region, under reasonably sheltered conditions but facing the open ocean, one finds in the upper half of the intertidal a distinct barnacle zone consisting mainly of small, closely-packed honeycomb barnacles, *Chamaesipho columna* (pl. 7 and fig. 5). Completely covering the surface of the rock in some areas, *Chamaesipho* may occur in concentrations of up to 12,500 individuals to the square foot and may form a broad, white, intertidal band, clearly visible even at a distance. In more exposed conditions the two surf barnacles *Catophragmus polymerus* and *Tetraclita rosea* (pl. 8) may partially or completely replace the smaller honeycomb barnacle. The presence of *Catophragmus*, with its numerous accessory plates, and the conspicuous, pinkish *Tetraclita* in numbers, indicates exposure to broken water and spray, and marks a rock slope as potentially dangerous for fishermen or marine biologists. Plate 8 shows two small specimens of another high intertidal barnacle, *Chthamalus antennatus*, in addition to the large surf barnacles. Partly obscured in the upper left by an empty *Tetraclita* and wedged between two *Tetraclita* on the right, the smaller *Chthamalus* can be recognised by the distinctive pattern of sutures between the main shell plates. The photograph also shows several molluscs characteristic of this upper intertidal barnacle zone. *Morula*, with black nodules on a pale background, is right of centre; *Melanerita*,

Plate 8. Surf barnacles, *Catophragmus polymerus* (centre) and *Tetraclita rosea*, x2.

Fig. 7. Pigmented shell of imperial purple barnacle, *Balanus imperator,* viewed through base, x2.

absolutely characteristic of this species. Another large barnacle found in this zone, though usually under more sheltered conditions, is the imperial purple barnacle, *Balanus imperator* (fig. 7). Its shell may appear pale externally but the plates themselves are deep purple throughout their internal structure. Broken fragments with this distinctive colouring can be seen in sand and shell grit near rocky areas wherever this species grows abundantly.

Several other acorn barnacles occur on rocky shores, in estuaries or as fouling growths on wharf-piles and ships in the Australian area. One of these, the small and insignificant-looking *Elminius modestus,* occurs naturally in New Zealand, but since the Second World War it has appeared in European waters and is now spreading rapidly as a major fouling organism. It is assumed that it was carried to the Northern Hemisphere attached to ships.

black with worn patches, is just above; *Melaraphe,* small and blueish-purple, is left of centre, and several limpets, including a large *Cellana* at top right, are scattered between the barnacles.

In the lower half of the intertidal, below the barnacle zone already described and under the same reasonably sheltered conditions, there is usually a characteristic zone of encrusting, white, calcareous tubes secreted by the polychaete worm, *Galeolaria caespitosa.* Low in this zone, and often extending into the still lower and partially subtidal cunjevoi zone, are scattered specimens of the giant rock barnacle, *Balanus nigrescens* (pl. 9 and fig. 8). This is one of the largest barnacles found in Australian waters and individuals may reach a height of 2½ inches. The cerulean blue cuticular edging exposed in the aperture as the shell valves open is

Fig. 8. Rock barnacle, *Balanus nigrescens,* with cirri just completing feeding stroke, x2½.

Plate 9. Open valves of large rock barnacle, *Balanus nigrescens,* x3. 

ISOPODA—Shore Slaters and Fish Lice

THE REMAINDER OF the groups discussed in this book are all 'higher Crustacea' or Malacostraca. These animals have a fixed number of body segments, nineteen, and a matching number of appendages, nineteen, excluding the eyes. They also have their trunk limbs divided into two sets, thoracic and abdominal, the former of eight pairs (maxillipeds and legs) and the latter of six pairs (swimmerets and uropods, fig. 16).

The isopods, or 'equal-footed' malacostracans, are typically depressed, in other words flattened from above down, and are often oval in outline when viewed from above. Their thoracic legs are usually simple, rather than chelate, or nipper-bearing, and look alike, while their eyes are not stalked but sessile on the sides of the head. The almost cosmopolitan garden slater, or smooth wood louse, *Porcellio laevis,* so common under stones or timber in damp situations, is a very familiar example of this group. It was introduced to Australia from Europe soon after colonisation and is found now over almost the entire continent. Many species of isopods, both free-living and parasitic, occur in the sea; some occur in freshwaters and some, such as the garden slater, are truly terrestrial.

Another familiar isopod is the shore slater, *Ligia,* with at least two species in Australia, *Ligia exotica* (fig. 9) and *Ligia australiensis*. Shore slaters are inhabitants of rather sheltered rocky shores and are capable of very rapid movement when disturbed.

Fig. 9. Shore slater, *Ligia exotica,* a fast-moving intertidal scavenger, x2. 

Plate 10. An intertidal pill bug, *Zuzara venosa*, male, x13.

They are active intertidal scavengers and may at times temporarily invade boat slips, dockyards and waterfront warehouses a few feet above high tide mark. *Ligia* can be so common on some parts of the South Australian shoreline that scores of individuals are disturbed when a single stone is turned over.

Damage by timber-boring isopods to wooden structures in the sea is a problem of major economic importance in Australian harbours. The gribble, *Limnoria tripunctata,* about an eighth of an inch long, and two pill bugs, *Sphaeroma quoyana* and *Sphaeroma terebrans,* both about half an inch long, are the main species involved in New South Wales. They bore into wharf piles, wooden oyster stakes, submerged timber of all kinds and even sandstone and the lead covering of cables. A large, untreated wharf pile has been known to collapse, under the combined attack of these three species and the amphipod *Chelura terebrans,* in less than two years. Modern treatment can bring this problem under control, but will probably never eradicate it.

There are many marine isopods of the general pill bug shape and form that do not bore into timber or stone. *Zuzara venosa* (pl. 10) for instance is a common intertidal pill bug in temperate Australian waters. Males of this species have a distinctive, backwardly-directed horn. Other marine isopods are known for their carnivorous habits; sea lice of the genus *Cirolana,* in areas where they are common, can clean dead fish down to skeletons overnight. *Cirolana* can nip the legs of bathers and surf fishermen in shallow water causing irritating little wounds.

Parasitic isopods are well known to fishermen under various names, for instance fish lice, tongue-biters and fish doctors. Some are found clinging to the outside of fish, often near the tail, as for example the impressively-coloured, tropical *Rocinela* species shown in Plate 11, while others lodge under the gill covers or in the mouth of larger fish. Some, like the leatherjacket louse, *Ourozeuktes owenii* (fig. 10), even burrow through the sides of fish and remain partly within the body cavity and partly projecting out through the skin. Others again, the bopyrids, are parasites in the gill chambers of decapod crustaceans.

Fig. 10. The leatherjacket louse, *Ourozeuktes owenii,* a parasite which burrows into the body cavity of fishes, x1½.

Plate 11. A strikingly-coloured, tropical fish louse, *Rocinela* species, x8.

AMPHIPODA—Sandhoppers and Skeleton Shrimps

CLOSELY ALLIED to the isopods are the amphipods, or 'double-footed' malacostracans. In contrast to the former group, amphipods are typically compressed, or flattened from side to side, and one or more pairs of thoracic legs are chelate or nipper-bearing. Like the isopods they have sessile rather than stalked eyes. As can be seen in figure 11, the last three pairs of legs are opposed to the others, in other words the direction of closing is reversed in these posterior legs and this feature is the basis for the name 'double-footed' The amphipods are an abundant group, not only in the sea, where free-living, commensal and rare parasitic forms occur, but also in freshwaters and, in some parts of the world, on the land.

Amphipods have been described poetically as having 'many twinkling feet' and as carrying about with them almost as many tools as a plumber. A whole series of different appendages are modified in the one animal for feeling, biting, sorting, holding, brushing, swimming, jumping etc. In the family Talitridae, which includes all terrestrial amphipods, or leafmould hoppers (fig. 11), and most high intertidal shore sandhoppers, the last three pairs of abdominal limbs are all modified for jumping and this is the characteristic method of progression in this group. A leafmould hopper or a sandhopper balances on the third to last pair of legs while turning the abdomen under the body so that the ends of uropods and telson (for terms see fig. 16) press on to the ground. The last two pairs of legs are held parallel to but do not touch the ground. When the abdomen is suddenly straightened out the animal is propelled into the air. On landing, the abdominal limbs and the last two pairs of legs are used as shock absorbers.

True terrestrial amphipods, as distinct from high intertidal or supralittoral species, have a tropical and Southern Hemisphere distribution and live almost entirely in forest leafmould. They are not known to occur naturally in Europe, northern continental Asia, North, or South America. Where present, as in Australian tropical and temperate rain forests, they have been recorded in concentrations as high as 4,000 individuals per square yard. Under these circumstances they can form a very conspicuous part of the cryptozoic, or leafmould, fauna, both because of their numbers and because of their habit of jumping when the litter is disturbed. They eat fallen leaves and thus play a major part in the disintegration of leaf litter in Australian rain forests. The common leafmould hopper of south-eastern Australia and Tasmania, *Talitrus sylvaticus* (fig. 11), occurs also in New Zealand and has been accidentally introduced to the British Isles. A closely allied species, *Talitrus pacificus,* ranges from eastern and western Australia to East Africa and the islands of the South Pacific, with accidental introductions to Hawaii, California, Louisiana and, most recently, the Azores.

The amphipods as a group are richly represented

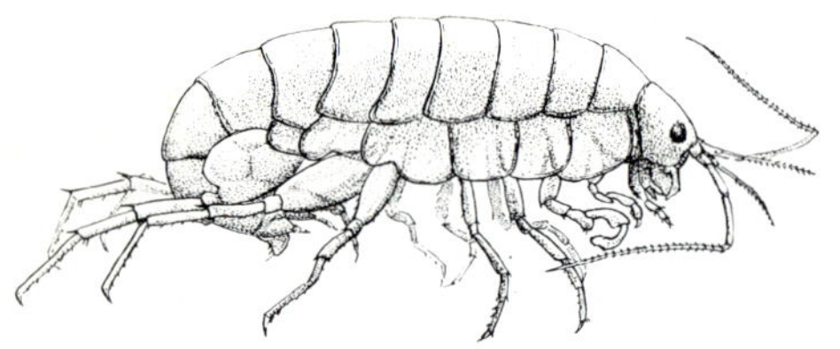

Fig. 11. Leafmould hopper, *Talitrus sylvaticus,* a terrestrial amphipod, x12.

Plate 12. An intertidal amphipod, *Amphithoe* species, x12.

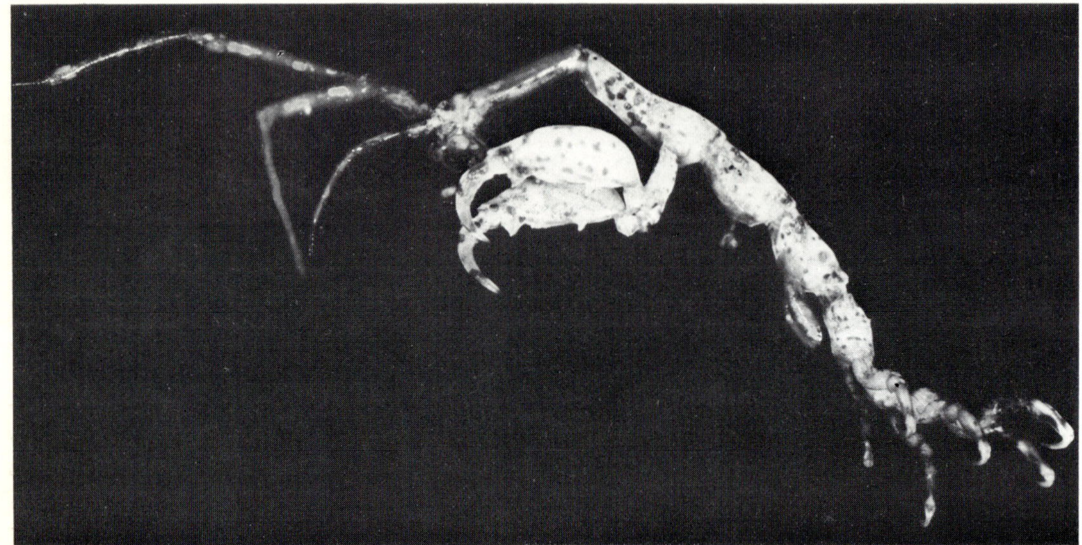

Fig. 12. Skeleton shrimp, *Caprella equilibra*, an elongate amphipod from intertidal seaweed, x6.

on Australian seashores. Numerous species live among tide-line litter, under stones or in rock pools; some live among seaweeds and other marine growths in both shallow and deep waters; some shelter in sponges, sea-squirts and other invertebrates; some burrow into wood, while many are true planktonic, wide-ranging, oceanic forms. Examples include *Orchestia chiliensis,* a common shore sandhopper in New South Wales, and the mottled, pink-eyed *Amphithoe* species (pl. 12), which ranges widely in temperate-Australian intertidal rock pools. Plate 13 shows two closely-allied, intertidal amphipods, *Ceradocus sellickensis* and *Ceradocus ramsayi.* They often occur together among seaweeds, but can be readily distinguished by different toothing on the enlarged subchelate nippers.

Two very distinctive, but atypical, amphipod subgroups are the skeleton shrimps, or caprellids (fig. 12), and the whale lice, or cyamids. The former are slender, elongated creatures, small in size, which live in large numbers on intertidal and shallow water seaweeds. They assume the same colouring as the weed and move slowly amongst it in a distinctive, looping manner, bringing the posterior part of the body forward and arching the back, then moving the anterior part of the body forward to obtain a new grip. Note the greatly enlarged nippers of the male *Caprella equilibra* in Figure 12. Whale lice look superficially like isopods, as they are depressed rather than compressed, but they agree in all essential features with the amphipods. They live as external parasites on the skin of whales, clinging by a series of curved claws, and are one of the very few

Plate 13. Intertidal amphipods, *Ceradocus sellickensis (top)*, *Ceradocus ramsayi (bottom)*, x15.

amphipod subgroups which can be regarded as truly parasitic.

The relatively large and richly coloured amphipod, *Paraleucothoe novaehollandiae* (pl. 14), lives in the branchial baskets of ascidian sea-squirts. This is an excellent example of what is termed a commensal rather than a parasitic relationship. Commensalism, with a literal meaning of 'eating together at the same table', is the state where one living thing lives on or in another animal or plant at no expense to its host. The commensal shares the food of the host or feeds on matter the host does not require, but does not harm or directly benefit the host. Other commensal crustaceans mentioned in this book are the black-lipped pearl shrimp, *Conchodytes* (pl. 23), the tropical krill shrimp, *Galathea* (pl. 31), and the harlequin crab, *Lissocarcinus* (fig. 45).

All ascidians have a large sac-like pharynx perforated by many rows of pharyngeal slits. The beating of cilia draws a current of water through the mouth, into the pharynx, through the pharyngeal gill slits and into the atrium, which is the cavity surrounding the pharynx. Food particles are trapped in the pharynx by a close meshwork of filaments across the gill slits, and the water leaves by an opening from the atrium to the outside of the animal. The complex meshwork of slits and filaments making up the wall of the pharynx is called a branchial basket. Large, solitary sea-squirts around the temperate Australian coastline, such as the cunjevoi *(Pyura stolonifera)*, the sea-tulip *(Pyura pachydermatina)* and the subtidal, tan-coloured *Pyura spinifera* (fig. 13), often contain commensal *Paraleucothoe*. The extraordinary and dramatic effect of plate 14 has been achieved by photographing a living *Paraleucothoe* in place inside the complex branchial basket of a cut-open *Pyura spinifera*. The sinuous yellow folds of the inside of the branchial basket can be seen clearly in

the plate. Note the sharp-clawed walking legs of the amphipod gripping the meshwork. It is assumed that these amphipods, after entering a *Pyura* mouth when young, grow to become prisoners within its pharyngeal walls.

Fig. 13. Subtidal ascidian, *Pyura spinifera*, x½. With the related cunjevoi and sea-tulip, this species often contains commensal amphipods *(opposite plate)*.

  Plate 14. Commensal amphipod, *Paraleucothoe novaehollandiae*, in branchial basket of ascidian, x10.

Fig. 14. A tropical mantis shrimp, *Gonodactylus graphurus*, x2.

STOMATOPODA—Mantis Shrimps

THE STOMATOPODS, or 'stomach-footed' malacostracans, are a small group of crustaceans (40-odd species in Australia), fairly uniform in appearance. They live in intertidal sandflat burrows, intertidal rock or coral crevices and subtidally on all types of bottom. Stomatopods are characterised by unusual and powerful raptorial claws on the second thoracic legs. These have a coarsely-toothed end segment which folds forwards like the blade of a pocket knife on to the preceding segment. The resemblance of this claw to that of a praying mantis insect is the reason for the name 'mantis shrimp'. Species of the common subtidal genus *Squilla* are often taken in commercial prawn hauls and are sometimes called, with little real justification, 'prawn-killers'.

DECAPODA—Prawns, Crayfish and Crabs

THE REMAINDER OF this book is concerned with the large and varied group of decapod, or 'ten-footed', malacostracans. So diverse are the members of this group (defined on page 6) that they are divided into three major subgroups: macrurans or shrimp-like decapods, anomurans or decapods with reduced tails, and brachyurans or true crabs. Included in this assemblage of different forms (totally about 1,200 in Australia) are commercial crustaceans such as penaeid prawns, marine crayfish and swimming crabs, as well as numerous shrimps, freshwater crayfish, ghost nippers, hermit crabs, spider crabs and other crabs of all types and sizes.

Plate 15. Cooked king prawns, *Penaeus plebejus,* in shop display. 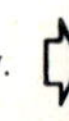

MACRURA—Shrimp-like Decapods

Commercial prawns (Family *Penaeidae*)

IN ADDITION TO ten pairs of similarly-sized walking legs (pls. 16 and 25) or one pair of enlarged nippers and four pairs of walking legs (pls. 18 and 29), the shrimp-like decapods, or macrurans, have an elongated and subcylindrical body with a long, well developed abdomen terminating in a tail-fan. A rostrum, which may be large and blade-like (fig. 16), projects forwards between the eyes and both pairs of antennae have long, slender flagellae. The pleopods, or abdominal limbs, are well developed in all macrurans, including crayfish, and are used as swimming organs in shrimps and prawns. The eyes are on movable stalks in almost all decapods (pls. 21 and 49), although a few aberrant species are virtually eyeless and in the alpheid snapping shrimps (pl. 18) the eyes are covered over by a forward projection of the carapace.

In Australia, commercial species of the family Penaeidae are universally called 'prawns'. These can be distinguished from most other prawns and shrimps (termed, in contrast, carideans) by their possession of nippers on the first three pairs of legs. In addition, penaeid prawns shed their eggs directly into the sea after fertilisation. All other decapods carry their eggs about with them until they hatch. Many different penaeids are taken commercially in Australian waters but about thirteen species make up the majority of the commercial catch. In Western Australia these include western king prawns, in northern Australia (including the Gulf of Carpentaria) banana and tiger prawns, and in eastern Australia school, king, and greasy-back prawns.

The eastern king prawn, *Penaeus plebejus* (pls. 15 and 16), can be taken as a fairly typical commercial species. It ranges from northern Queensland along the coast of eastern Australia to Victoria, and it grows to a length of 10 inches or more. Its life cycle begins in autumn with an egg lying freely on the bottom at the offshore spawning grounds in a depth of about 50 to 80 fathoms. The egg hatches in about 12 to 18 hours. The first larval stage, the nauplius, is attracted by light and moves up towards the surface. A series of naupliar, protozoeal and mysis larval stages are passed through planktonically and after about three weeks postlarval prawns about half an inch long return to the bottom. At this stage juveniles move inshore and enter coastal inlets and estuaries. In these areas of low and fluctuating salinity, growth is rapid and nine to eleven months later the kings have reached a size of about 3½ inches but are still sexually immature. They then migrate seawards in dense schools, called 'spawning runs', usually with an outgoing tide at night. Growth

Fig. 15. Sorting a commercial prawn catch on a trawler at night.

Plate 16. Living, commercial king prawns, *Penaeus plebejus*, x1½.

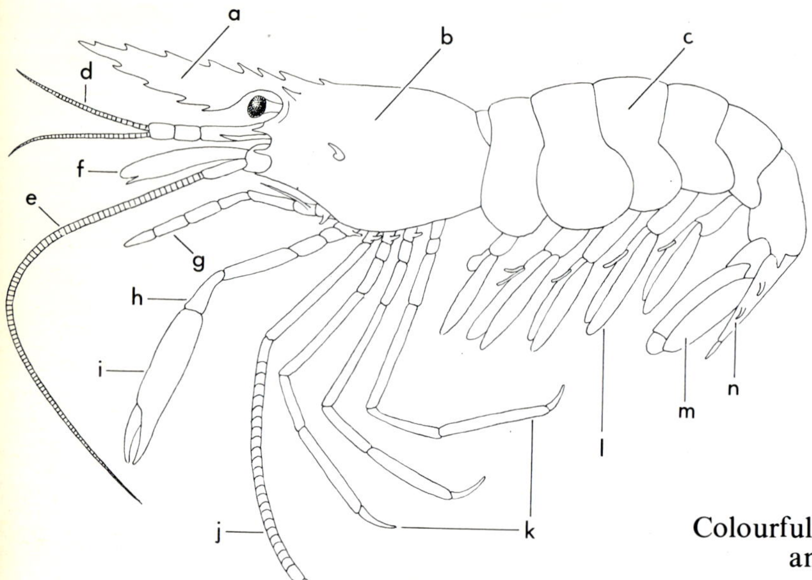

Fig. 16. A generalised prawn with principal external features labelled. *a*, rostrum; *b*, carapace; *c*, 6-segmented abdomen; *d*, double-branched 1st antenna; *e*, 2nd antenna; *f*, scale or scaphocerite; *g*, 3rd maxilliped; *h*, wrist of 1st leg; *i*, palm of 1st leg, forming with the two fingers a hand or chela; *j*, multiarticulate wrist of 2nd leg; *k*, 3rd to 5th legs; *l*, swimmerets or pleopods; *m*, uropods; *n*, telson, forming with uropods a tail fan.

continues in offshore waters and the prawns mature and spawn at a body length of 5 inches or more. Most king prawns die after spawning, when one year old, but some live into their second year and continue to grow. These apparently do not spawn a second time and no mature kings ever return to inshore waters.

All other commercial penaeids, except the greentail or greasy-back, *Metapenaeus bennettae*, are offshore breeders like the king prawn, but not all enter estuaries in their juvenile phase. Greentails complete their entire life history in estuarine waters, leaving only during extensive flooding.

Colourful caridean shrimps and prawns

THE GENERAL NAMES 'shrimp' and 'prawn' are used rather indiscriminately for the lightly built, swimming macrurans, though 'shrimp' is often applied to smaller forms lacking a prominently-projecting rostrum. In contrast, the heavily built, often massively clawed, crawling macrurans are called 'crayfish' or 'lobsters' (differences on page 54).

Unlike the rather uniform penaeids, the caridean shrimps and prawns show great diversity in shape, size and colour. Figure 16 shows a generalised caridean, but great variation in the form and appearance of the many appendages is possible. The rostrum can be long and slender (pl. 22), movable (fig. 18) or all but absent (pl. 18 *bottom*); the eyes can be large and round (pl. 17), small (pl. 23) or covered by the carapace (pl. 18 *bottom*). The third maxillipeds, used as extra mouth-parts to manipulate food, are usually leg-like in appearance, but

Plate 17. The hinge-beak prawn, *Rhynchocinetes rugulosus*, x6.

may become greatly elongated at maturity (pl. 18 *top,* projecting forward from prawn) or flattened into a series of foliaceous plates (fig. 22, between chelae). The first and second legs are usually chelate, but the chelae, or hands, can be enlarged (fig. 17), elongated (fig. 19) or flattened (fig. 22). The third to fifth legs are usually used as supporting and walking structures, but may bear plumose side-branches used for swimming (pl. 22, second from top). The combined carapace and abdomen, or body, of the caridean is usually streamlined and smooth (fig. 20) but may be short and dumpy (pl. 23); it may bear long spines (fig. 23) or even tufts of coloured hair-like setae (page 5).

The caridean macrurans are divided into some twenty-two families, at least fifteen of which are represented in the Australian fauna. Many of these are rare or obscure groups but common or interesting members of six different families will be discussed in this and the following chapter.

The hinge-beak prawn of eastern Australia, *Rhynchocinetes rugulosus* (pls. 17, 19 and back cover), is one of the most colourful and, in one aspect, one of the most unusual of all carideans. Members of this family, the Rhynchocinetidae, all belong to the one genus and all have in common the

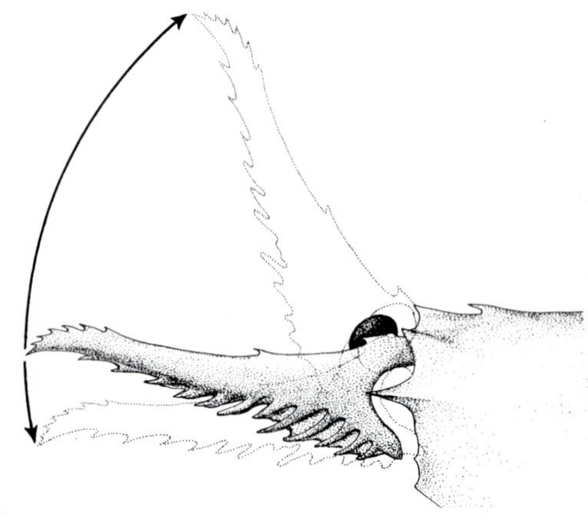

Fig. 18. Hinged rostrum of *Rhynchocinetes* prawn showing extent of movement possible in vertical plane.

extraordinary feature of the rostrum being hinged and movable in a vertical plane (fig. 18). A movable rostrum is otherwise virtually unknown in the macrurans and the name 'hinge-beak' for these animals is singularly appropriate and distinctive. *Rhynchocinetes rugulosus* is known from the Indonesian Archipelago to Japan and Hawaii, and down the east coast of Australia to the Victorian border. Another species, *Rhynchocinetes australis,* occurs in South Australian waters. The eastern Australian hinge-beak is found on rocky shores under stones at low tide level, and in rock pools among seaweeds; it is also well known to skindivers in rock crevices and small caves, and has been trawled from depths down to 40 fathoms. Its complex colour pattern is unmistakable and is shown in detail in the accompanying plates.

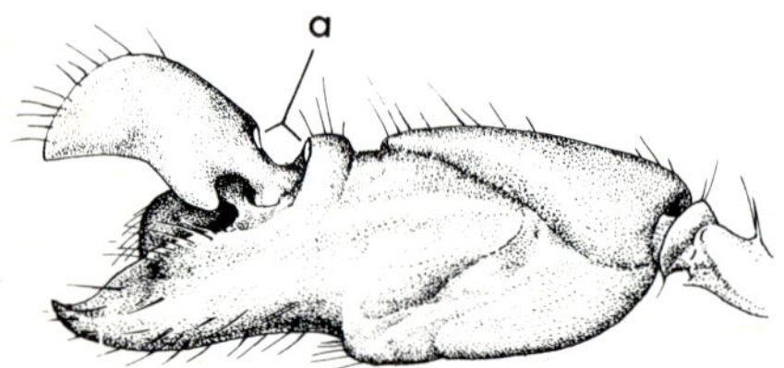

Fig. 17. Large hand of snapping shrimp, *Alpheus,* partly open to show projecting peg on inner margin of movable finger. *a* indicates matching adhesive surfaces which make contact when hand is fully open.

Plate 18. Intertidal shrimp, *Alope orientalis (top)* x4; snapping shrimp, *Alpheus* species *(bottom)* x3.

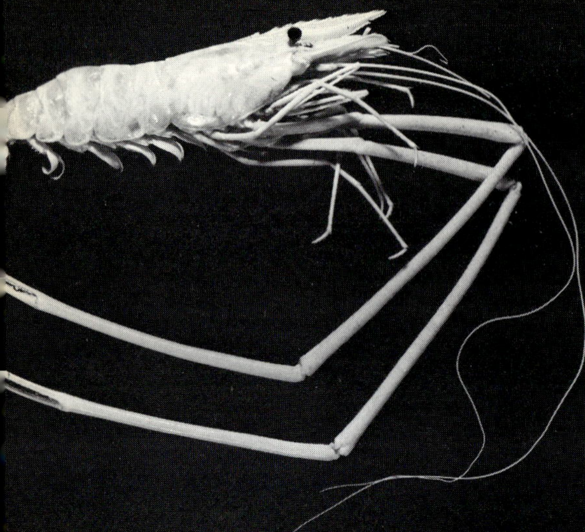

Fig. 19. The long-armed prawn, *Macrobrachium novaehollandiae*, from eastern Australian, brackish waters, x⅔.

A colour pattern such as this is bilaterally symmetrical. In other words, each side of the prawn is the mirror image of the other, and, except for a certain amount of individual variation, is constant for all members of a species. The spots and bands of colour are made up of aggregations of very small colour-cells, called chromatophores, arranged on the surface of the body immediately under the transparent cuticle. In crustaceans each one of these microscopic organs consists of a central area with many, branched, hollow processes radiating out in several directions. The pigment, in the form of ultra-fine granules, is carried in a fluid inside the chromatophore, and this fluid can flow outwards until the whole chromatophore and its branches appear full, or it can be retracted and concentrated in the central area. Thus the chromatophore is spoken of as being in an expanded or contracted state, though it is the pigment which moves since the hollow branches are permanent. Some chromato-

phores are compound structures consisting of two or more differently coloured, but unmixed, pigment cells. The central areas are discrete but juxtaposed and the individual sets of branches radiate and intertwine without joining. The expansion and contraction of crustacean chromatophores, and the subsequent changes of colour involved, are under the control of glandular secretions, or hormones, produced in the eye-stalks and distributed by the blood system. Crustacean colour changes are not rapid but are made to match gross changes of background and differing amounts of illumination. Thus *Rhynchocinetes,* after an hour or more on dark green seaweed, appears darker and greener than usual, while on pale red sponge it would appear paler and redder. There is also a night to day rhythm and the prawn is paler at night and darker by day. In plate 19 many of the different chromatophore types present on the hinge-beak prawn can be seen: simple bluish-white chromatophores between bands of simple brownish-red chromatophores, and scattered, compound red and white chromatophores on the carapace at *top left;* simple red chromatophores clumped on the rostrum at *top right;* bluish-white chromatophores bunched in spots on the carapace, and white chromatophores in bands around the legs at *bottom left,* and simple yellow chromatophores forming long bands on the abdomen at *bottom right.* The authors consider this to be one of the most dramatic and complex displays of colour on a small scale in the animal kingdom.

Snapping shrimps of the family Alpheidae are more often heard than seen by visitors to seashore, swamp and mudflat. There are many forms in Australia, especially in tropical reef waters, but a red-banded *Alpheus* species from the eastern Australian, rocky, intertidal zone is illustrated here (pl. 18 *bottom*) as a large, but typical, example of this group. Snapping, or pistol, shrimps are capable

Plate 19. Chromatophores of hinge-beak prawn, *top r* and *l* x45, *bottom r* and *l* x6. 

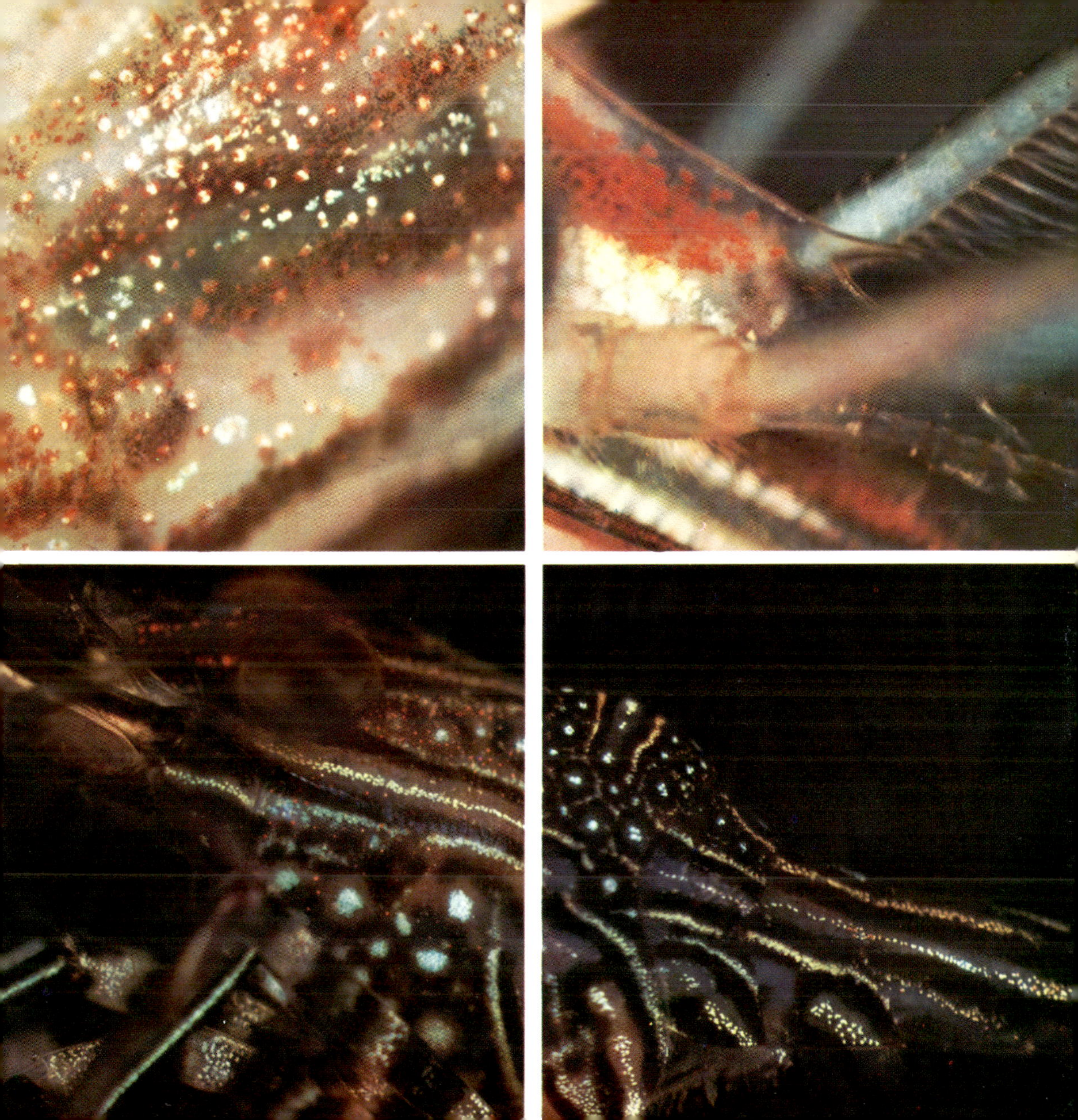

of producing a loud and sharp crack, very much like the crack of breaking glass, with their massively-enlarged hand. The first pair of legs are modified into strong nippers in these shrimps and one member of the pair is invariably much larger than the other. This big hand has a peg-like projection on

Fig. 21. The wide-ranging zebra shrimp, *Gnathophyllum american-um*. A small specimen with variant colour pattern, x6. (Compare *opposite plate*).

the movable finger of the nipper which fits into a socket on the fixed finger (fig. 17). The sudden closing of these fingers under strong muscular tension, like the triggered release of a gun hammer, produces a snapping sound as the peg goes into the socket and the heavily calcified tips of the fingers meet. This sound-producing mechanism, and the associated jet of water displaced by the peg, appears to be used in both offence and defence by these shrimps.

The large family Palaemonidae includes freshwater, intertidal and commensal forms; it is well represented in and around the Australian continent. The common freshwater and estuarine genus *Macrobrachium* has many species in this area and includes some giants in northern tropical rivers. The long-armed prawn, *Macrobrachium novaehollandiae* (fig. 19), an eastern Australian estuarine form, is often taken with similarly-sized, penaeid prawns in coastal lakes and lagoons. The extraordinarily

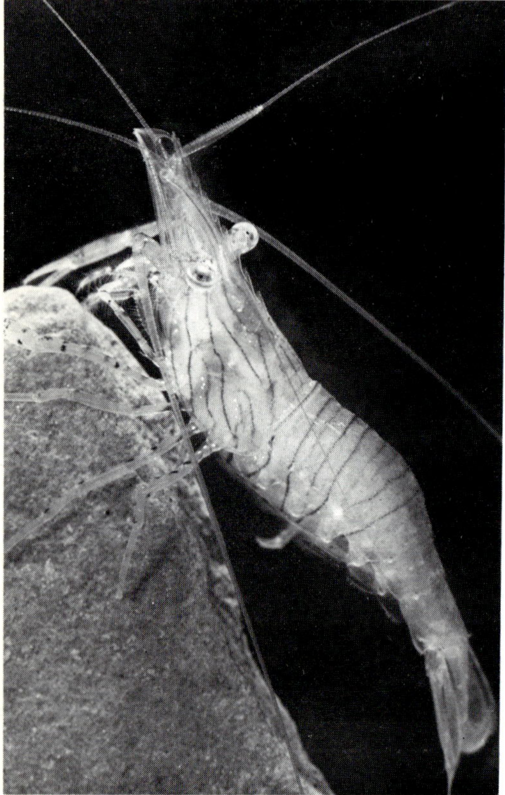

Fig. 20. A rock-pool shrimp, *Palaemon litoreus*, from eastern and southern Australia, x4.

Plate 20. The zebra shrimp, *Gnathophyllum americanum*, x12.

elongated second legs, over twice the length of the body when fully developed, usually arouse interest in commercial fishermen and have given rise to fanciful stories of malformation and monstrous growth in 'prawns' brought about by 'weather', 'atomic-bomb testing' or other natural or man-made phenomena! *Palaemon litoreus* (fig. 20) is one of several rock-pool palaemonids around our coast. Its transverse red stripes on the abdomen serve to distinguish it from the more common and often associated *Palaemon serenus* which has longitudinally-aligned red spots and flecks on the abdomen. *Conchodytes,* a commensal palaemonid, is discussed on page 50.

A colourful, widespread, coral-reef prawn, *Saron marmoratus* (page 5), is a representative of another family, the Hippolytidae, or hump-backed prawns. The irregular patterning of small circles on the body, the long, banded first legs and the bunches of red setae are all characteristic of this species, while the prominent hump in the abdomen indicates the family. Smaller hippolytids are common on our temperate shores, notably *Alope orientalis* (pl. 18 *top*) under stones and in rock pools, and *Hippolyte ventricosa* in seaweed.

The small and little known family, Gnathophyllidae, is represented in Australia by two unusual and rare (or perhaps more accurately described as 'seldom seen') genera, *Gnathophyllum* and *Hymenocera*. The zebra shrimp, *Gnathophyllum americanum* (pl. 20 and fig. 21), is virtually circumtropical as it has been found in Florida, the West Indies, the eastern Atlantic, the Indian Ocean area, the Indonesian Archipelago, Japan, Oceania and in eastern Australia as far south as Sydney. This dramatically-patterned little shrimp has been seen in at least two colour phases in our area. Most commonly it has broad black body bands and orange and black leg bands as in the colour plate, but nearly transparent specimens occur with narrow body bands and unbanded legs (fig. 21).

The elegant coral shrimp, *Hymenocera picta* (pl. 21 and fig. 22), ranges from East Africa and the Red Sea through the Indonesian Archipelago to Hawaii, New Caledonia and the Great Barrier Reef. When seen, the bold colour pattern and the plate-like, lamellate chelae readily identify this caridean. A pair of these shrimps under observation by Dr Rene Catala in the Noumea Aquarium, New Caledonia, performed a complex courtship 'dance' in slow motion. The smaller male appeared to follow the rhythm and movement of the female and this co-ordinated behaviour was highlighted in Catala's colour film *'Carnival under the Sea.'*

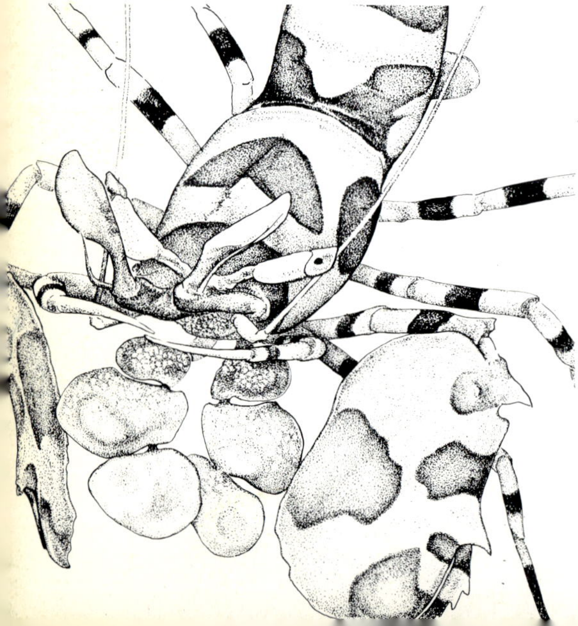

Fig. 22. Front view of *Hymenocera picta,* showing enlarged and flattened chelae on 2nd legs, x4.

Plate 21. The elegant coral shrimp, *Hymenocera picta,* x4. *Photo E. Slater.*

Fig. 23. Artist's representation of deep-sea prawns, *Oplophorus* species, producing momentary and brilliant 'luminous clouds' of light.

Deep-sea, commensal and cleaning shrimps

VISIBLE LIGHT PENETRATES about 1,500 feet down in open ocean waters. Below this depth is a region of perpetual darkness and there, under the lighted surface layers and above the colder bottom waters, is the realm of the bathypelagic animals. This midwater zone is sometimes called the 'black fish—red prawn' zone after the typical colours of the two dominant animal groups present. Some of the free-swimming shrimps and prawns living there are partially red and partially transparent, but the majority are entirely bright red. Common forms distributed all over the oceans of the world and found around Australia in the midwaters of the Tasman Sea, the Southern Ocean and the Indian Ocean include carideans of the family Oplophoridae, in particular species of the genera *Acanthephyra* (pl. 22, first two from top) and *Oplophorus* (fig. 23), as well as species of the penaeid and penaeid-like genera *Gennadas* (pl. 22, bottom) and *Sergestes* (pl. 22, third from top).

As well as being scarlet in colour, some of these deep-sea prawns can produce light, apparently at will, by several different methods. This light is a cold light or a luminescence, that is light produced with almost no loss of energy as heat. Some prawns

Plate 22. Deep-sea macrurans, *Acanthephyra, Sergestes, Gennadas,* x1½. *Photo P. Ralph.*

have external light organs called photophores (often lens-bearing), others have internal light organs such as modified areas of the liver, and some produce light in the form of a 'luminous cloud''. The latter is a discharge, from glands near the mouth, of luminous matter producing a momentary but brilliant flash of light (fig. 23). Animal cold light, as well as that produced by some bacteria, is known as bioluminescence, though the term phosphorescence is often loosely used for this phenomenon. The chemistry of bioluminescence is not fully understood, but in most cases it appears to be due to the reaction of luciferin with oxygen and water in the presence of an enzyme, luciferase. Extra energy is given off as light, a mere by-product in this unusual reaction.

The striking and almost unearthly nature of 'luminous cloud' bioluminescence was witnessed by one of the authors at sea off New Zealand in 1956. One morning before dawn a midwater net was brought up from 300 fathoms. A four-inch *Oplophorus novaezeelandiae* was taken out alive and undamaged and placed in seawater. Almost immediately the prawn discharged a vivid cloud of luminous matter from near the mouth. The light clearly illuminated the contents of the bucket, but faded rapidly and in a few moments was no longer visible. When the prawn was taken from the water a few minutes later it repeated this phenomenon. Luminous material was seen to originate near the mouth, then to flow over the author's wet hand on to the deck.

The palaemonid shrimp *Conchodytes meleagrinea* (pl. 23) is a commensal in the black-lip pearl oyster, *Pinctada margaritifera* (fig. 24), of tropical reef waters. *Conchodytes* ranges widely in the Indian and Pacific Ocean region and has been found from East Africa and the Red Sea to Hawaii and Oceania. In Australia it is known from north-west Australia,

Fig. 24. Partly open black-lip pearl oyster, *Pinctada margaritifera*, x½, host for commensal shrimp *Conchodytes*.

   Plate 23. Commensal shrimp, *Conchodytes meleagrinea*, pair *(left)*, in *Pinctada (right)*, x4.

around the northern coast to southern Queensland and occurs only in living black-lip pearl oysters. The left half of plate 23 shows a pair of these shrimps removed from the mantle cavity of a living oyster and photographed against a background of coral sand. The larger of the two is the female and the dark egg-mass carried under the abdomen is showing through the semi-transparent body. The red mass within the carapace is the mature ovary. The male, though smaller than the female in body size, has relatively larger chelae and very distinct red chromatophores. The right half of the plate shows a female shrimp in an opened black-lip pearl clinging to the edge of one of the oyster's gill folds. The threads at the top are the byssus, or attachment threads, of the oyster.

The banded coral shrimp, *Stenopus hispidus* (pl. 24 and fig. 25), is a colourful and extraordinary animal found throughout the tropical waters of the Indian and Pacific Oceans, and in the West Indies. It is not a caridean but belongs to a small group of macrurans (the stenopodideans) which have nippers on the first three pairs of legs, as do penaeids, but carry their eggs under the abdomen, as do carideans. *Stenopus* lives in coral or rock crevices and in underwater caves from low tide level right down through the depths usually visited by skindivers. It is known to occur in Australia from about North West Cape in Western Australia, around the northern and north-eastern coasts, south to Shellharbour in southern New South Wales. It is now widely known that the banded coral shrimp has the unusual habit of cleaning fish, a form of behaviour shown by only a few tropical shrimps and some tropical fish. Pairs of *Stenopus* have been observed in coral crevices with their feelers projecting and displayed in the sunlight. Fish are attracted to these waving white antennae (normally six branches to each animal) and remain still while the shrimp picks at parasites, injured tissue and fungal growths on their bodies and fins with its two pairs of small chelae. The shrimp does not use its large banded chelae for cleaning, nor does it leave its crevice to clean, but merely reaches out towards the fish which appear to congregate around known *Stenopus* cleaning sites. These fascinating shrimps form long-lasting pair associations, have been seen to perform court-ship 'dances' and, in one pair at least, the small male spent long periods 'saddle' riding on the back of the larger female (fig. 25).

Fig. 25. Pair of *Stenopus hispidus* with small male 'saddle' riding on female, x1; note interlaced antennae. *Photo F. G. Myers.*

Plate 24. The banded coral shrimp, *Stenopus hispidus*, x2. *Photo F. G. Myers.*

Marine crayfish
(Family *Palinuridae*) and allies

THE TERMS 'lobster' and 'crayfish' are used haphazardly in 'Australia for the large, edible, marine, macrurous crustaceans. The question as to which of the two names is more appropriate is often a source of doubt and argument. There are in fact no lobsters fished commercially in the Southern Hemisphere. The large commercial crustaceans fished in Australian waters belong to an entirely different group of decapods from the North Atlantic lobsters. Furthermore, they do not possess the lobster's characteristic large and powerful grasping claws on the first pair of legs.

The Australian animals should be referred to as 'marine crayfish' to avoid confusion with the equally well known 'freshwater crayfish'. The latter possess a pair of big claws and thus would appear at first sight to be closer to the true northern lobster than to our southern crayfish. However, they belong to another and different grouping again (page 60). The French use the term *'langouste'* to distinguish clawless marine crayfish from the European lobster or *'homard'*; freshwater crayfish are known as *'ecrevisse'*.

Marine crayfish belong to the family Palinuridae. The two largest Australian palinurids are fished off the eastern and southern coasts. These are the eastern or common crayfish, *Jasus verreauxii* (fig. 27), and the southern crayfish, *Jasus novaehollandiae* (long known under the name *Jasus lalandii*). The first is plentiful in temperate waters along the New South Wales coast, while the second is particularly abundant in the cooler waters of the Tasmanian—Victorian—South Australian region. Giant propor-

Fig. 26. The painted crayfish, *Panulirus versicolor,* common in northern and Barrier Reef waters, x¼. 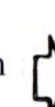

Plate 25. A coral crayfish, *Panulirus longipes,* x½. *Photo K. Gillett.*

tions are reached by the smooth-tailed, eastern crayfish, which can weigh as much as 17 pounds, be 3 feet in length (excluding antennae) and measure up to 11 inches across the widest part of the body. The usual weight of large specimens sold commercially is from 4 to 6 pounds. The red-coloured, southern crayfish, with its characteristic pavement patterning on the abdomen, can reach 11½ pounds in weight.

In Western Australia another major commercial palinurid is the western crayfish, *Panulirus cygnus*. The area fished for this species extends from Bunbury to Shark Bay in inshore and offshore waters and includes the reefs of the Houtman Abrolhos off Geraldton. This single species of crayfish supports the richest Australian commercial fishery. In 1968-69 the production was valued at about $18 million. This represented approximately 65 per cent of the total value of crayfish production in Australia and about 30 per cent of the total value of Australian edible marine products. The western crayfishery supplies frozen tails for the export market, sold under the name 'rock lobster', and whole crayfish for local consumption.

Several other colourful marine crayfish live in Australian tropical waters. They are difficult to capture as they seldom enter crayfish pots and are not fished commercially to any extent in this area. The commonest species in northern Australia is probably the vividly-patterned painted crayfish, *Panulirus versicolor* (fig. 26). The black markings on the carapace and the narrow black and white stripes across the green abdomen immediately distinguish this widespread species from other tropical crayfish. It is known off the north-western and northern coasts and in Barrier Reef waters. The ornate crayfish, *Panulirus ornatus* (page 7), has much the same distribution, but on the east coast has been found as far south as central New South Wales. Another species, *Panulirus longipes* (pl. 25), one of the so-called 'coral' crayfish, is widespread in the Indian and Pacific Oceans and is known in Australia between Heron Island, at the southern end of the Barrier Reef, and northern N.S.W.

Fig. 27. The eastern crayfish, *Jasus verreauxii*, important commercially in New South Wales; a young specimen, x1.

Plate 26. Rare reef lobster, *Enoplometopus occidentalis*, x1½. *Photo D. Henderson.*

Fig. 28. Red flapjack or shovel-nosed crayfish, *Arctides antipodarum,* from shallow south-eastern waters, x½.

The reef lobster, *Enoplometopus occidentalis* (pl. 26), is a rare and unusual macruran belonging to the same family as the true North Atlantic lobster (Nephropsidae). It has been found in widely scattered localities from the east African mainland to Hawaii and is now known in Australian waters. During night dives on the reef edge at Heron Island, Queensland, some years ago, several of these small lobsters were seen and brought to the surface. They were photographed but no specimens have been available for detailed examination.

Another family closely related to the palinurid marine crayfish is the Scyllaridae containing the flapjacks, or shovel-nosed crayfish, of temperate and tropical waters. These are clawless macrurans with rather flattened bodies and antennae enlarged and modified into rounded plates. Figure 28 shows the south-eastern shovel-nosed crayfish, *Arctides antipodarum,* occasionally seen by skindivers in shallow rocky waters. Another scyllarid is the 'Balmain bug', *Ibacus peronii* (pl. 27), smaller than *Arctides* and immediately distinguished by its serrated antennal plates. The 'Balmain bug' of New South Wales and the similar-looking 'Moreton Bay bug', *Thenus orientalis,* of Queensland and northern Australia, are both taken with prawn trawls and sold under a variety of names including 'sand lobster'.

Both palinurid and scyllarid crayfish have a most unusual, flattened and transparent larval stage in their life history, called the phyllosoma (fig. 29). This is a planktonic form which may last for six months or more. The phyllosoma grows by a series of moults and may travel far in ocean currents.

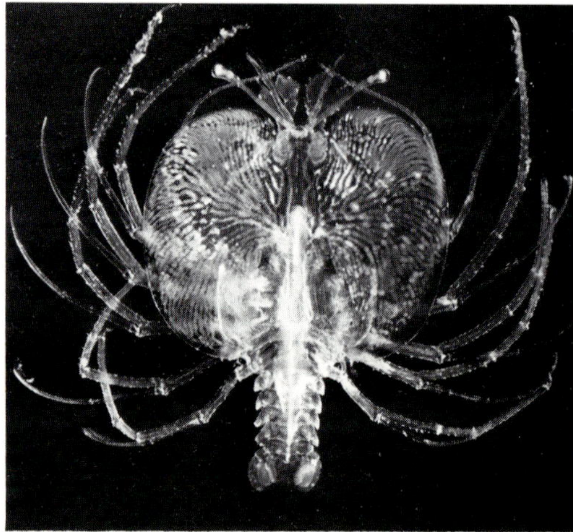

Fig. 29. Planktonic phyllosoma larva of scyllarid crayfish, probably *Ibacus peronii,* x1½.

Plate 27. 'Balmain bug' or sand crayfish, *Ibacus peronii*, x¾.

Freshwater crayfish
(Family *Parastacidae*)

THE FRESWATER CRAYFISH fauna of Australia is large and varied. There are over ninety species in nine genera on the mainland and in Tasmania. They range in size from inch-long adult midgets, such as *Tenuibranchiurus* of southern Queensland, to the giant Tasmanian freshwater crayfish, *Astacopsis gouldi,* which reaches 14 pounds or more in weight. These two species are in fact the smallest and the largest freshwater crayfishes in the world.

One of the interesting points about world freshwater crayfish distribution is its discontinuity. The northern freshwater crayfish, or astacids, are found in Europe, the northern Asiatic mainland and Japan, and in North America. The southern freshwater forms, or parastacids, are found in Madagascar, Australia, New Guinea, New Zealand and South America. There are no crayfish in the freshwaters of Central America, south-east Asia, India or the whole of the African continent.

The two commonest and most widely distributed parastacid genera in Australia are the spiny freshwater crayfish, *Euastacus,* and the smooth freshwater crayfish, *Cherax.* Species of *Euastacus* are found throughout the cooler areas of south-eastern South Australia, Victoria and southern New South Wales. They occur also in the mountain streams of northern New South Wales and southern Queensland, with one isolated species from mountain streams in northern Queensland. Species of *Cherax* have a wide distribution in Australia. They are found in south-western Australia (local names 'marron' and 'gilgie'), eastern South Australia,

Fig. 30. Murray River crayfish, *Euastacus armatus,* widespread in the Murray, Darling and Murrumbidgee drainage area, x¼.

Plate 28. Freshwater crayfish, *Euastacus australasiensis*, from coastal NSW, x1½.

Victoria, New South Wales, Queensland, coastal northern Australia and New Guinea, but not in Tasmania or in coastal, central New South Wales, including the Sydney area.

Three species of spiny freshwater crayfish are illustrated here. The Murray River crayfish, *Euastacus armatus* (fig. 30), is a large species with very distinctive white claws and large white spines on the tail. It grows to a weight of at least 6 pounds and is found in the Murray, Darling and Murrumbidgee River systems. *Euastacus australasiensis* (pl. 28) and *Euastacus spinifera* (pl. 29), are both found in the Sydney area. The former, usually red in colour, occurs in small 'soaks' and springs on coastal headlands to the south of Sydney Harbour, while the latter, usually dark greenish with red markings, is widespread in small streams, especially to the north of the harbour. The smooth freshwater crayfish, or yabby, *Cherax destructor* (fig. 31), is one of the widespread inland species of south-eastern Australia. It causes damage to farm dams, retaining walls and irrigation channels by perforating and undermining these structures with burrows. Specimens can survive the temporary drying up of such water bodies by sealing themselves in deep burrows terminating in water-filled chambers.

The small, so-called land crayfishes of south-eastern Australia and Tasmania, species of the genus *Engaeus,* live in family groups in deep burrows, often at a considerable distance from surface water. They are the only crayfishes with this form of social behaviour. Some of the species build chimneys of mud above the entrance to their burrow, others burrow in lawns and orchards causing damage to the roots of plants.

Fig. 31. Smooth freshwater crayfish, or yabby, *Cherax destructor,* common in inland streams, dams and irrigation ditches, x⅔. 

Plate 29. Freshwater crayfish, *Euastacus spinifera,* from Sydney area, x¾.

Ghost nippers, lobster krill and half crabs

DECAPOD CRUSTACEANS are conveniently divided into three groups. These are the macrurans or straight-bodied shrimps, prawns and crayfish, the anomurans or 'decapods with reduced tails', and the brachyurans or true crabs, with their miniature abdomens folded forwards under the carapace. The anomurans occupy a rather intermediate position between the other two groups. Their bodies are either crab-like (pl. 32) or shrimp-like (pl. 30) in general shape, and their abdomens variously modified. The latter are either bent more or less beneath the carapace (pl. 31), or spirally-twisted and soft (pl. 35), or, if extended, they are feebly-calcified with a narrow attachment to the carapace (pl. 30).

Shrimp-like anomurans burrow in sand and mud or live in burrow-like cavities in rocks, corals and sponges. They can be distinguished from macrurans by their reduced abdomen, reduced eyes and absence of antennal scale (f in fig. 16). Crab-like anomurans are either free-living or commensal rather than burrowing forms, and usually appear superficially eight-legged. They can be distinguished from brachyurans by their long antennae (compare pl. 32 with fig. 46), by the presence of a tail fan and by their habit of holding the slender and reduced fifth pair of legs bent up against the carapace (pl. 31) or tucked away in the gill chambers under the carapace (fig. 37). Brachyurans in contrast have short antennae, no tail fan and fully functional and extended fifth legs. Hermit crabs, the 'typical' anomurans in the minds of most people, have an asymmetrical and spirally-twisted abdomen which is normally kept hidden away in an empty gastropod shell.

An exceedingly abundant, eastern Australian, shrimp-like anomuran is the ghost nipper, or

Fig. 32. Using 'yabby pump' to obtain *Callianassa australiensis* from estuarine sand flat.

Plate 30. Male ghost nipper, or marine yabby, *Callianassa australiensis*. x1½.

or nipper, of the first pair of legs is always asymmetrically enlarged and flattened. This hand becomes very large in adult males and the shape of the curved and gaping fingers is characteristic for this species. In Queensland these *Callianassa* are called 'yabbies', but the word can cause confusion as it is widely used in other states for the smooth freshwater crayfishes of the genus *Cherax* (see preceding chapter). Other names used are bait, burrowing, pink and sand shrimp.

Ghost nippers occur from northern Queensland to Victoria in estuaries or in the lower reaches of coastal rivers. They are most abundant on intertidal sand banks or on mixed sand and mud flats (fig. 55). A permanent burrow is excavated by a *Callianassa* early in its juvenile life and is extended both vertically and into side galleries as the crustacean grows. Burrowing is done with the nippers of the first pair of legs plus the two following pairs of legs. The latter are used to scoop the sand forwards into a 'basket' formed between the nippers and the mouthparts. The sand in this 'basket' may be sifted for organic food through the hairs on the mouth parts or carried to the entrance of the burrow and pushed outside. As the burrow is only a little wider than the animal's body, *Callianassa* is faced with a 'one-way traffic' problem. This is overcome by the excavation of wider chambers at intervals. At these chambers the animal can reverse direction by somersaulting. The burrow of a large *Callianassa* may reach a depth of at least 30 inches and have up to three surface openings. In densely crowded ghost nipper beds there may be up to 800 openings per square yard of surface and adjacent burrows often interconnect.

Surface openings are easily recognised in sand flats, and ghost nippers are dug up or pumped out by fishermen collecting for bait. A manually-operated, commercially-made 'yabby pump' is

Fig. 33. Southern lobster krill, *Munida subrugosa*, occasionally swarms in Bass Strait, x2.

marine yabby, *Callianassa australiensis* (pl. 30). This pale, pinkish-white creature, about three to four inches in length, is highly sought after by fishermen for bait. It lives in burrows in intertidal, estuarine sand flats and rarely leaves its burrow under normal conditions. The name 'ghost nipper' alludes to the pale, almost translucent appearance of the animal, coupled with the fact that one chela,

Plate 31. Elegant squat lobster, *Galathea elegans*, a tropical commensal, x6.

shown in figure 32. The open end of the metal tube is placed over a burrow opening and the pump is pushed into the sand. It is then extracted, removing a core of sand and emptied with the plunger. The pump is reinserted into the hole and suction applied with the plunger. Water, sand and ghost nippers are drawn into the pump and can be emptied out at the surface.

Another thalassinid, or shrimp-like anomuran, is the seldom-seen, pink *Laomedia healyi* (frontispiece). This is a deep burrower both in eastern Australian mangrove swamps and in the mud banks of subtidal channels draining such swamps. Note the small eyes, the symmetrically enlarged chelae of the first legs and the longitudinal hingeline along the side of the carapace. The carapace flap below the hingeline covers the gill chamber and is capable of in-and-out movement. This 'panting' assists in the circulation of water within the poorly oxygenated burrow.

Lobster krill of the family Galatheidae are anomurans with elongated, forwardly-extending nippers and bodies which are longer than wide. But for the abdomen bent forward under the carapace, and for the reduced, hunched up fifth legs, galatheids look superficially like small lobsters. The red, subtidal *Munida subrugosa* (fig. 33) is known from the Bass Strait area and is occasionally found on the bottom in great swarms. It apparently does not have a pelagic juvenile phase as does the closely related southern New Zealand *Munida gregaria*. Swarming galatheids in many parts of the world form a major source of food for large fish, seals and whales. The elegant squat lobster, *Galathea elegans* (pl. 31), is a strikingly-patterned tropical form found commensally associated with reef crinoids. Light longitudinal stripes are characteristic of a number of small crustacean commensals, both carids and galatheids, which live on crinoids. ·

Another group of anomurans are the crab-like Porcellanidae, commonly called porcelain or half crabs. They have enlarged nippers which extend laterally and then fold forwards on to themselves. Free-living intertidal forms move sideways like crabs and scuttle rapidly under cover when disturbed. A characteristic Barrier Reef species, *Petrolisthes lamarckii*, is shown in plate 32. It is most abundant on beach rock areas under loose coral boulders and ranges widely through the tropical Indian and Pacific Oceans.

The hairy stone crab, *Lomis hirta* (fig. 34), is a crustacean enigma. It is a crab-like anomuran with long antennae, a symmetrical abdomen and the fifth legs hidden from sight. Its relationships are obscure and *Lomis* cannot be classified readily in any anomuran subgroup. It is restricted to southern Australia and ranges from Tasmania to south-western Australia.

Fig. 34. Hairy stone crab, *Lomis hirta,* an uniquely-Australian, southern, intertidal anomuran, x2.

Plate 32. A tropical half crab, *Petrolisthes lamarckii,* x4. *Photo K. Gillett.*

Hermit crabs
(Family *Paguridae*) and allies

T HE ASSOCIATION BETWEEN hermit crabs and mollusc shells is well known. This extraordinary dependence on the dead shell of another animal belonging to an entirely different group is linked with the nature of the hermit crab's soft, asymmetrical and spirally twisted abdomen.

Hermit crabs of the family Paguridae have an elongate and more or less cylindrical body made up of a carapace, which is hard in front but soft behind, and a relatively large abdomen (pl. 35). The first and last segments of the latter are small and calcified but the other four are soft and show little trace of segmentation. The gonads, liver and renal organs, usually confined to the thorax in other decapods, extend into this soft membraneous abdomen. The tail fan is specially modified for holding the pagurid securely inside a gastropod shell. The left appendage is larger than the right and both have roughened, file-like, gripping surfaces which are pressed against the inside of the spiral shell. The abdominal pleopods are reduced in number and often missing entirely on the right side. The first pair of legs are chelate and usually massive; often one is much larger than the other. The second and third pairs of legs are long and strong (pl. 34), while the fourth and fifth are reduced and sometimes weakly chelate.

All crustaceans grow in size by moulting many times during their life. After the old cuticle is shed, the new one is soft and the crustacean is extremely vulnerable to predatory animals until this has hardened. During the first few days after moulting, many crustaceans hide themselves away under stones, in sand or in crevices. Hermit crabs, however, are constantly protected by their acquired

Fig. 35. Small, south-east Australian, intertidal hermit crabs, *Pagurus lacertosus,* removed from shells, one carrying eggs, x2.

shell and have no need to hide during or after moulting. Because of their use of dead shells, pagurids have the constant problem of finding a new shell every time they outgrow the one they occupy.

When the time comes for a hermit to change its home it will investigate a number of different shells as it moves about. It touches any shell encountered and if there is no response to this disturbance it will probe inside carefully with its antennae. A large number of shell features are important in determining whether or not the hermit crab will select the shell for occupation. These include its ability to be moved, its surface texture, external shape, nature of its aperture, internal size and its weight. A pagurid is able to distinguish very effectively between shells of the same species but of differing weights, and to select a shell appropriate to its own size. If satisfied

Plate 33. Common eastern Australian, intertidal hermit crab, *Pagurus sinuatus,* x6.

The two commonest hermit crabs on the open coastal, rocky reefs of south-eastern Australia are the small and hairy-handed *Pagurus sinuatus*, which is red and favours turban shells as shown in plate 33, and the very small and abundant *Pagurus lacertosus* (fig. 35). The latter may occur in hundreds under stones near low water mark. It favours peri-winkles and small kelp shells, but the preference depends on the relative abundance of such shells in different localities. These two pagurids are restricted Australian forms, while *Clibanarius virescens* (pl. 34) with its characteristic white eye-ring and yellow leg markings is, in contrast, a very widely distributed hermit crab. It has been recorded from East Africa and the Red Sea, through the Indonesian Archipelago to Japan, New Caledonia and Fiji. In Australia it is known from the western, northern and eastern coasts as far south as Sydney.

The most unusual Australian pagurid is the stone-inhabiting *Cancellus typus* (fig. 36) from southern and south-eastern Australia. It is an exception to the almost universal rule that pagurids shelter in mollusc shells. *Cancellus* lives in short, more or less straight burrows (apparently self-made) in rounded, water-worn pebbles. The stones selected are generally coarse sandstones, very granular in texture, and a little bigger than the pagurid itself. This hermit lives subtidally and has been taken by dredges and trawls, but in some localities in the Sydney area it has been found in numbers just below low water mark during spring tides. *Cancellus* drags its stone home around under water quite actively and when disturbed withdraws completely into the burrow and closes the circular entrance with its closely fitting and externally flattened chelae and second legs.

The brilliantly-coloured *Dardanus megistos* (pl. 35 *bottom*) is the largest and most spectacular hermit crab known in the Barrier Reef area. Many examples grow to a length of 8 inches or more,

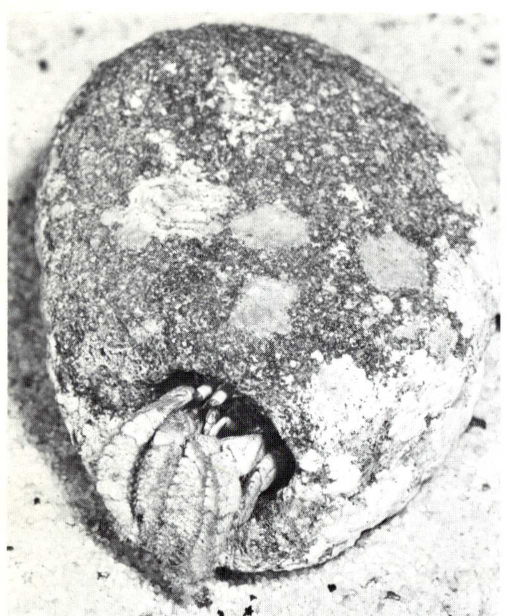

Fig. 36. 'A stone that walks'; the shallow water, stone-inhabiting miner hermit crab, *Cancellus typus,* from southern and eastern Australia, x1.

that the shell is unoccupied and large enough, the hermit crab places itself alongside the empty shell, hastily withdraws its soft abdomen from its old home and slowly and carefully inserts it into the new shell. After selecting a new home the pagurid continues to investigate other shells and the amount of time spent doing this depends on the suitability of its recently acquired shell. The characteristic behaviour associated with the selection and entry into shells does not depend on previous experience; it is fully developed when the juvenile hermit crab selects and enters its first home.

Plate 34. A widespread, equal-handed hermit crab, *Clibanarius virescens*, x8.

including the powerful nippers. The red body has an overall scattered pattern of very distinct white spots, black-ringed on the carapace and legs, but tending not to be so ringed on the abdomen. *Dardanus megistos* is one of the few pagurids to be pigmented over the whole body surface including the abdomen and reduced abdominal appendages. Most pagurids lack colour pattern, or even surface pigment of any kind on the hind end of the body normally hidden within a shell. This red and white-spotted *Dardanus* has a wide distribution in the tropical and temperate Indo-Pacific, and ranges from East Africa and the Red Sea in the west to Hawaii and Tahiti in the east, and from Japan in the north to Australia in the south. On the Australian east coast it extends well down into temperate New South Wales.

Land hermit crabs of the family Coenobitidae are in general a tropical group poorly represented in the Australian area. Some small, brown *Coenobita* species occur in northern Australia and the bright red *Coenobita perlatus* (pl. 35 *top*) is abundant on cays in the Coral Sea between Queensland and New Caledonia. Though found widely throughout the tropical Indo-Pacific from the Red Sea to Samoa, this active scavenger is seldom seen on the islands in the Barrier Reef area. It does occur, however, on cays in the Swain Reefs, on islands of the Capricorn and Bunker groups and on Lady Elliott Island, all at the extreme southern end of the Reef proper. These virtually terrestrial decapods have branchial chambers richly supplied with blood vessels as well as gills, and can use atmospheric oxygen directly. They can live for many months out of water but usually return to the shore quite regularly to wet themselves, change shells and release larvae from hatching eggs.

The subantarctic stone crab, *Lithodes murrayi* (fig. 37), is the only representative in the Australian area of the mainly northern hemisphere family Lithodidae. Including important commercial species in other areas, these usually large forms are true anomurans, related to the hermit crabs (female abdomens are asymmetrically twisted to the right), but are not shell-inhabiting. *Lithodes murrayi* lives in the shallow waters around Macquarie Island, 800 miles south of Tasmania, and empty carapaces and legs are commonly found washed ashore on the beaches of this isolated subantarctic island.

Fig. 37. Subantarctic stone crab, *Lithodes murrayi*, from Macquarie Island, south of Tasmania, x⅓.

Plate 35. Land hermit crabs, *Coenobita perlatus (top)*, x½; *photo K. Gillett*. White-spotted hermit, *Dardanus megistos (bottom r and l)*, x½.

Frog crabs (Family *Raninidae*)

THE LAST OF the three groups of decapods to be considered are the brachyurans or true crabs. The name means 'short tail', and this single feature, coupled with the lack of a tail fan, clearly distinguishes this large and varied group from all other decapods. Brachyurans comprise more than 50 per cent of all decapod crustaceans known from

Fig. 38. The harp crab or smooth frog crab, *Lyreidus tridentatus*, an offshore burrower from both eastern and western Australia, x2.

Australian waters, and the 650 or more species recorded from the area are classified into some twenty-three different families. A selection of these families and species is discussed in the following pages.

The long-bodied frog crabs, or raninids, are very different in general shape from all other brachyurans. They are burrowers in offshore and shallow water muds and sands, and are seldom seen unless collected by trawls or, in the case of the red frog crab, *Ranina ranina* (pl. 36), foul-hooked on fishing lines. Every summer several large specimens of *Ranina*, sometimes called spanner crabs from the shape of the large chelae, are brought in to the Australian Museum, Sydney, for identification. They have usually been taken in shallow water off central New South Wales by line fishermen. The colour, size and shape of the crab, as well as the blade-like digging feet, arouse great interest in the inquirer and it is often difficult to explain that the animal is a reasonably well known, widely distributed and edible crab in the tropical Indo-Pacific area and that it is quite regularly taken in Sydney waters. It ranges from East Africa and the Indonesian Archipelago to Japan and Hawaii. In Australia it has been recorded from the Abrolhos in Western Australia, through northern Australia and Barrier Reef waters, to as far south as southern New South Wales.

The much smaller harp or smooth frog crab, *Lyreidus tridentatus* (fig. 38), is found in deeper waters off both south-eastern and south-western Australia. Its distributional range around this continent is discontinuous as it has not been taken off northern or southern Australia. It also occurs in New Zealand, where it is sometimes called the Tojo crab, and in Japan and Hawaii.

Plate 36. The frog or spanner crab, *Ranina ranina*, x$\frac{2}{3}$.

Box and sand crabs
(Family *Calappidae*)

IN MOST CRABS a respiratory current of water is drawn into each gill chamber at the base of the cheliped and passes out near the mouth parts. Three groups of burrowing crabs show basic modifications in this water circulation system. In the raninids (page 76) water is drawn into the gill chambers between the first abdominal segment and the bases of the fifth legs, while in the calappids and in the leucosiids (page 82) water is drawn in at the front, through a channel below the eyes, and passes out at the front from beneath the elongate mouth parts.

There are several species of the box crab genus *Calappa* in Australian waters, all characterised by a wing-like expansion on each side of the carapace, under which the walking legs can be withdrawn and concealed. The chelipeds are very large and com-pressed and can be held close against the front of the body. The common box crab, *Calappa hepatica* (figs. 39 and 40), is a widespread, tropical Indo-Pacific species which, in Australia, ranges from the Abrolhos in Western Australia, through northern and eastern Australia, to central New South Wales. All box crabs have a large tooth on the movable finger of the right hand, fitting between two prom-inences on the fixed finger (fig. 39). This apparatus is used to break open gastropod shells piece by piece so that enclosed hermit crabs can be eaten.

Burrowing sand crabs of the genus *Matuta* occur in shallow waters all around the Australian continent, except for Victoria and Tasmania. They are strong swimmers and most species have a pair of prominent lateral spines. The double-lined sand crab, *Matuta planipes* (pl. 37), is known from the Indian Ocean, from north-western, northern and eastern Australia to northern N.S.W., and from Japan.

Fig. 39. Right chela of box crab, *Calappa hepatica*, a specialised 'tool' for opening gastropod shells, x2.

Fig. 40. Common box crab, *Calappa hepatica*, a widespread tropical species, x¾.

Plate 37. Sand crab, *Matuta planipes*, with digging and swimming legs, x2.

Sponge crabs (*Dromiidae*) and pebble crabs (*Leucosiidae*)

T HE DROMIIDS, or furry sponge crabs, are the least specialised of all brachyurans and are closely allied to the anomurans. Almost all dromiids carry about with them a piece of living sponge, ascidian (sea squirt), soft coral or shell held in position over their backs by the last two pairs of legs. To achieve this the fourth and fifth legs are reduced in

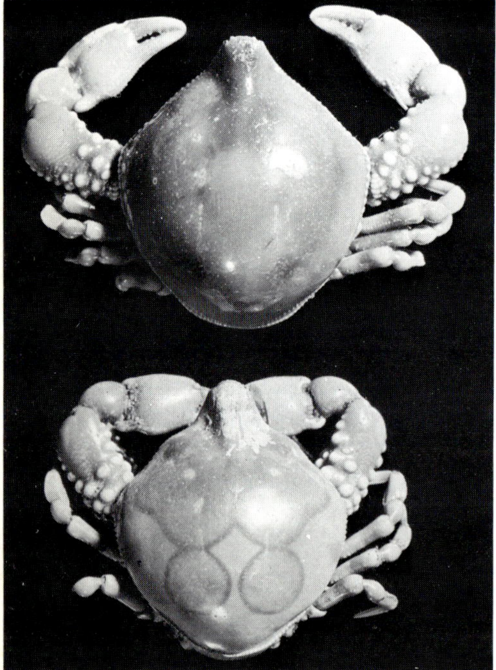

Fig. 41. Widespread, tropical pebble crab, *Leucosia anatum*, white-spotted form *(top)*, red-circled form *(bottom)*, x2.

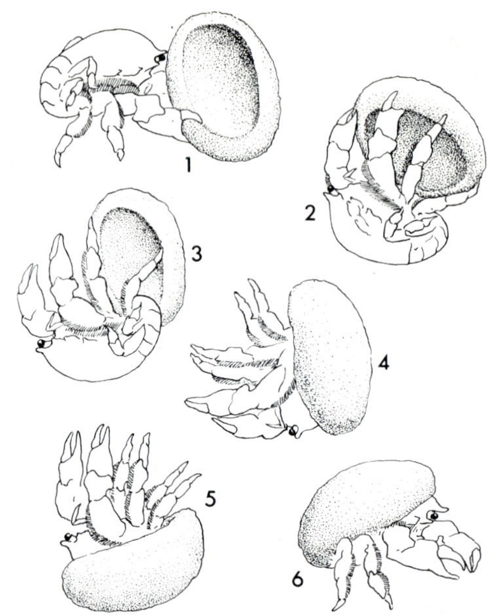

Fig. 42. A sponge crab putting on its own self-made sponge cap.

size, directed more or less upwards and are minutely chelate. There are several intertidal and shallow water species in Australia, the commonest being *Petalomera lateralis,* ranging from north-eastern Australia, south round the continent to north-western Australia. As with some other crabs, the eggs of this dromiid hatch directly into young crabs, instead of into free-swimming larvae, and these young spend some time sheltering under the abdomen of the mother. Plate 38 shows an intertidal *Dromidia* species with its self-made sponge cap over the back of the carapace. Dromiids deprived of their own caps will go to great lengths to retrieve them, but will cut and shape new caps if necessary. In putting on

Plate 38. Sponge crab, *Dromidia* species, carrying sponge cap, x7.

its own cap (fig. 42), a sponge crab may grasp the cap, roll over on its back, upend itself backwards into the cap and then right itself.

The pebble crabs, or leucosiids, are a characteristic little family typified in shape and appearance by the wide ranging, tropical, subtidal and shallow water species *Leucosia anatum* (fig. 41). This shiny, globose, Indo-Pacific pebble crab exists in several colour forms, and ranges in the Australian area from north-western Australia, through northern and north-eastern Australia, south to central New South Wales. It is normally taken by dredge from soft shallow bottoms. In contrast, the shape of the slender-legged, eastern Australian spindle crab, *Ixa inermis* (fig. 43), is very unusual for a leucosiid. Its arched carapace is grossly extended sideways by large, blunt-ended processes, which may in some specimens be unevenly shaped or sized.

Spider crabs (Families *Majidae* and *Hymenosomatidae*)

APART FROM THE furry and cap-carrying sponge crabs (pl. 38), there are four unequally-sized groups of crabs, each with its own more or less characteristic shape. There are the long-bodied raninids with the carapace longer than wide (pl. 36), there are the related calappids (pl. 37) and leucosiids (fig. 41) with the carapace more or less circular in outline (excepting wing-like expansions and enlarged lateral processes), there are the spider crabs (discussed in this chapter) with the carapace more or less triangular in outline, apex forwards, and there are all the other, so-called brachyrhynchous, crabs (covered in the following six chapters) with the carapace more or less square (pl. 47), oblong (pl. 51) or oval (pl. 42) in outline.

Fig. 43. The spindle crab, *Ixa inermis*, with immensely developed lateral carapace spines, x2.

Plate 39. Carapace of ornamented sea-toad, *Schizophrys dama*, x3.

One of the characteristic features of spider crabs of the family Majidae is the presence on the carapace and legs of special curled or 'hooked' hairs which aid in the attachment of animal and plant growths, especially seaweeds. These growing organisms are placed in position by the crab itself using its chelipeds which can reach the upper surface of the carapace. For this reason majid crabs are often called masking, seaweed, decorator or camouflage crabs. Some spider crabs deliberately place sea anemones on their backs and hold them carefully in position with a chela until the anemones have become attached to the carapace by their basal discs. This is a form of enforced commensalism as the sea anemone presumably obtains some food fragments missed by the majid, and the crab may gain some protection from the presence of the anemone and its stinging tentacles. Majids show colour perception, and have been observed to replace pink seaweed growing on a carapace with green seaweed, when taken from a patch of pink algae and placed in green surroundings. A fully decorated spider crab with silt and fibres among the various growths on the body is almost unrecognisable to the uninitiated and usually gives itself away by moving.

About ninety-five species of majid spider crabs are known from the Australian area. Most of the species in tropical Australia are widespread Indo-Pacific forms, while most temperate Australian species have discontinuous or restricted distributions. The ornamented sea-toad, *Schizophrys dama* (pl. 39), is known from south-east Asia and from Western and northern Australian shallow waters. A carapace on its own is illustrated here, completely denuded of growths and hairs. The symmetry and surface sculpturing of this magnificent natural structure with its paired rostral horns is shown in detail on this plate though it is normally completely obscured in life.

Fig. 44. Estuarine spider crab, *Halicarcinus australis;* a male with enlarged hands, x1½.

The hymenosomatid spider crabs are a small, rather poorly known family with their world distribution centred in Australia and New Zealand. They are little crabs, rarely reaching an inch across the carapace, and are sometimes called false spider crabs or, inappropriately, sea-spiders. A common, southern Australian, intertidal and shallow water species is the 'three-pronged' *Halicarcinus ovatus* (pl. 40). This tiny crab, usually associated with algae, has a minutely trifid rostrum and ranges from central New South Wales south and west to southern Western Australia. It occurs in a variety of dramatic, usually symmetrical, colour patterns. The related and similarly distributed *Halicarcinus australis* (fig. 44), grows to a larger size and occurs in estuarine and brackish conditions, while *Halicarcinus lacustris* is a freshwater form from southern Australia, New Zealand, Norfolk and Lord Howe Islands.

  Plate 40. Variation in an intertidal spider crab, *Halicarcinus ovatus, top* x8, *bottom r* and *l* x3.

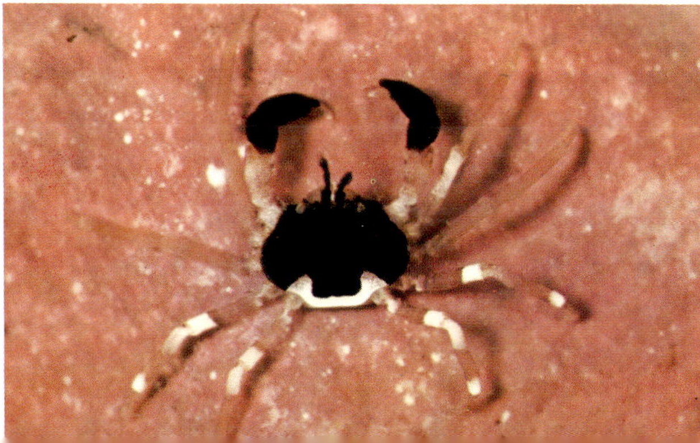
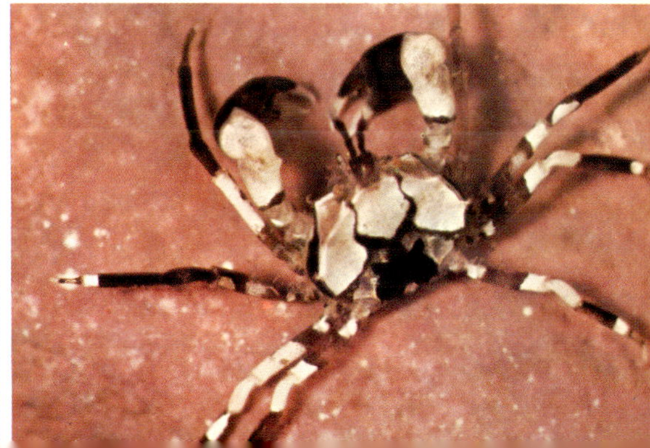

Swimming crabs
(Family *Portunidae*)

THE TWO STRIKING features which characterise most portunid crabs at first sight are the flattened fifth legs, forming 'swimming paddles', and the tendency for the carapace, especially the toothed anterolateral borders (*g* in fig. 46), to be broadened. The flattened legs in portunids, as in the sand crab *Matuta* (pl. 37), are used for both swimming and burrowing. Despite the family's common name, most swimming crabs are expert burrowers in sand and mud while some are only indifferent swimmers. Broadening of the carapace helps to streamline the body for swimming sideways, and the broadest portunids, the blood-spotted (pl. 41) and the blue (fig. 46) swimming crabs, appear to be the strongest swimmers.

Not all the eighty-odd portunids in Australian waters are free-living, and some do not have strongly flattened fifth legs. The parti-coloured harlequin crab, *Lissocarcinus orbicularis* (fig. 45), is a commensal portunid living either on the outer surface, among the oral tentacles, or within the gut of large tropical holothurians, or sea-cucumbers. It is a wide-ranging Indo-Pacific species and occurs in eastern Australia as far south as Moreton Bay, Queensland. The mated pair shown in figure 45 consist of a white male with a reddish-black pattern (above) and a reddish-black female with a white pattern (below). The 'positive' black on white of the male is the more usual colouring of this species, but the reverse pattern, the 'negative' white on black seen in the female is also known. These two patterns have not been seen together in a pair before.

The large and edible blue swimmer or bluey, *Portunus pelagicus* (fig. 46), is the only decapod crustacean, as far as is known, with a distribution extending completely around the Australian continent. This active and beautifully mottled crab grows to a width of seven inches or more and is fished commercially in Queensland, especially in Moreton Bay where it is called the sand crab. It is well known in the estuaries of New South Wales and Victoria. In South Australia it is sometimes called

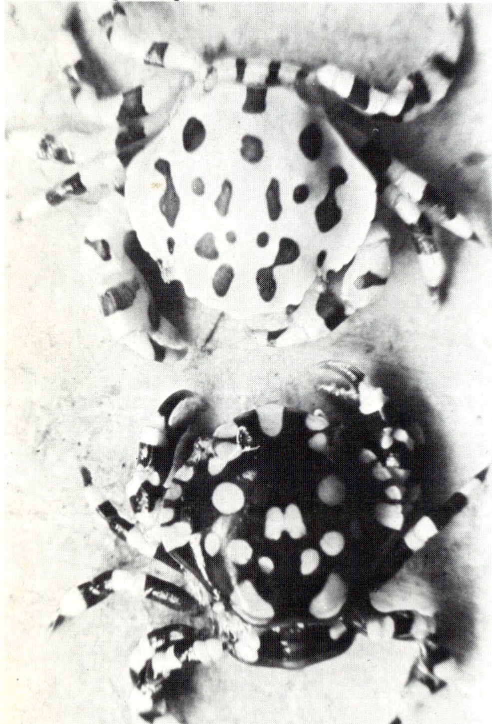

Fig. 45. Pair of harlequin crabs, *Lissocarcinus orbicularis*, commensal on tropical sea-cucumber; male above, female below, x2. *Photo Julie Booth.*

Plate 41. *(Top)* blood-spotted swimming crab, *Portunus sanguinolentus,* x½; *(bottom left)* detail of spots and barnacles, x1½; *(bottom right)* egg-mass, x½.

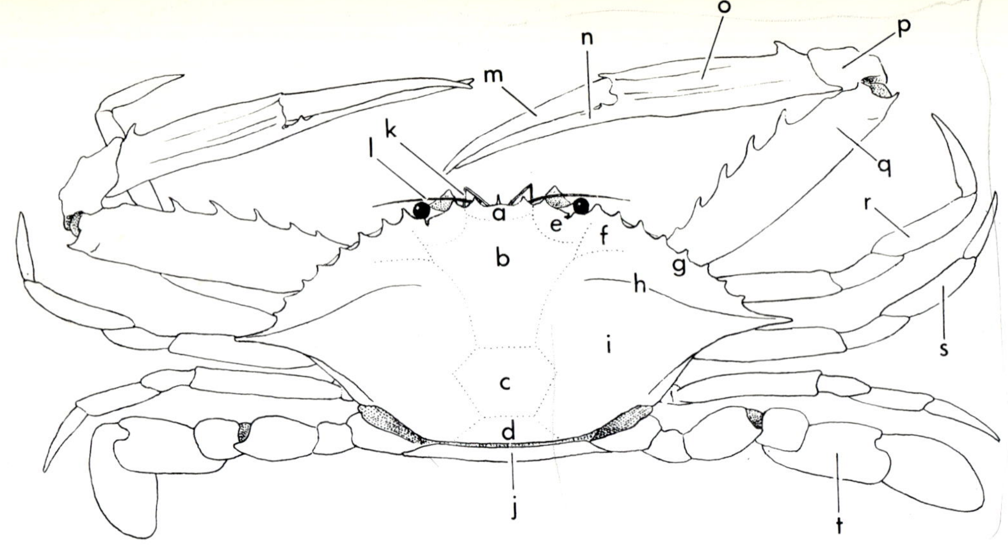

Fig. 46. The blue swimming crab, *Portunus pelagicus*, with principal external features labelled. *a–f*, regions of carapace; *a*, frontal; *b*, gastric; *c*, cardiac; *d*, intestinal; *e*, orbital; *f*, hepatic; *g*, anterolateral border and spines; *h*, epibranchial ridge; *i*, branchial region; *j*, abdomen folded under carapace; *k*, 1st antenna; *l*, 2nd antenna; *m*, movable finger of cheliped or 1st leg; *n*, fixed finger of cheliped; *o*, palm, forming with the two fingers a hand or chela; *p*, wrist of 1st leg; *q*, arm; *r*, 2nd leg (2nd–5th legs called walking legs); *s*, 3rd leg (or 2nd walking leg); *t*, 5th leg, in this family modified into swimming paddle.

the blue crab and regularly swarms into the shallow waters of St Vincent and Spencer Gulfs, where thousands are caught each year. Outside the Australian area it ranges from East Africa through the Indo-Pacific to Japan, Tahiti and occasionally to northern New Zealand. It has passed through the Suez Canal and is now fished commercially in the eastern Mediterranean. The somewhat smaller, blood-spotted swimmer, *Portunus sanguinolentus* (pl. 41), has a similar but not quite so extensive Australian distribution (it does not occur in south-eastern Australia) and ranges in the Indo-Pacific from East Africa to Hawaii. The three, white-bordered, red spots on the carapace immediately identify this edible species.

The massive and greenish-brown mud or mangrove crab, *Scylla serrata*, is another portunid of commercial importance in Australia. It is a widespread, tropical Indo-Pacific species living in large burrows in mangrove swamps. Elusive and nocturnal, this swimming crab is hunted and caught for food by a variety of methods in Queensland and northern New South Wales. Another common portunid is the bait-snatching surf crab, *Ovalipes australiensis* (pl. 42) of temperate Australia. This active, ocean beach crab ranges from southern Queensland south around the continent to Rottnest Island, Western Australia. The North Atlantic, edible shore crab (English name) or green crab (American), *Carcinus maenas*, is quite abundant on parts of the Victorian coastline, but its presence there is a mystery. Probably it was accidentally introduced by shipping, though it may be naturally antipodean.

Plate 42. The ocean-beach surf crab, *Ovalipes australiensis*, x1½.

Dark-fingered crabs
(Family *Xanthidae*)

IN ADDITION TO the short antennae and the reduced, forwardly-folded abdomen, brachyurans in general have a large and more or less flattened carapace covering the head and thorax from above and firmly fused in front to a bar-shaped plate lying below the antennae and in front of the mouth (see pl. 46). In this way a firm frame for the mouth cavity is formed. Within this frame the jaws and other mouth parts are covered in by the flattened third maxillipeds acting like a pair of folding doors. The first pair of legs, or chelipeds, bear strong, often relatively large, nippers or chelae, which may be sexually dimorphic in size. In such cases the stronger and more swollen chelipeds belong to the male. (Front jacket plate and fig. 56 show an extreme case of sexual dimorphism coupled with gross asymmetry in the male.)

The abdomen is relatively wider in a female crab than in a male. The narrow abdomen of the male, with only two pairs of styliform pleopods underneath, contrasts sharply with the broad abdomen of the female, which can be almost circular in outline in some species. The female abdomen bears at least four pairs of pleopods fringed with hair for the attachment of eggs. In many crabs the actual egg-mass is so large that only a portion is covered by the abdomen, but in some the eggs may be entirely concealed under and enclosed by the abdomen. Most crabs hatch into a planktonic larval stage called a zoea, with long carapace spines but few appendages. After numerous moults and a considerable increase in size the zoea enters the megalopa stage in which it looks like a miniature adult but does not have the abdomen tucked under the body. It then sinks to the bottom, moults again and emerges as a juvenile crab.

Brachyurans as a rule walk in a very distinctive manner, using their legs differently from all other crustaceans. The four pairs of walking legs, or in some crabs the first two or three pairs, are used for locomotion, while the chelipeds are reserved for grasping, feeding, fighting and other specialised activities. The walking legs of one side push and those of the other side pull so the crab walks or runs sideways. Those of the same side do not move together but alternately so that there is no halting of the synchronised movement. The speed achieved by such limb action can be amazing. Ghost crabs (page 102), for example, can move on sandy beaches at speeds of up to ten miles an hour. A complete exception to sideways walking is shown by the mictyrid soldier crabs (page 108), which all walk forwards.

Fig. 47. Giant Tasmanian deep-water crab, *Pseudocarcinus gigas*, reaching width of 16 inches and weight of 30 pounds, x¹⁄₁₆.

Plate 43. The shallow-water, spiny-legged crab, *Actaea peronii*, x4.

Fig. 48. The shawl crab, *Atergatis, floridus,* common and widespread in tropical Indo-Pacific, x1½

An English writer has suggested that the sideways movement of a crab can be demonstrated in the following way. Cross the two wrists side by side and secure together with an elastic band; hold the wrists well up from a table and place the finger tips on the table surface. Now let one set of fingers crawl while the other set pushes and keep up a continuous motion sideways without assistance from the arms.

The Xanthidae, or black-fingered crabs, are the largest family of decapod crustaceans in Australian waters. About 170 species in nearly 50 genera have been recorded from the area and the number is bound to increase as the tropical north is examined in more detail. Xanthids have their greatest diversity in numbers, shapes, colours and habits in tropical, especially coral reef, waters, but members of this family occur throughout the southern Australian area as well. The common name of the family is derived from the fact that in almost every xanthid the fingers of each hand, and sometimes part of the palm as well, are clearly, and usually very distinctly, dark in colour. In many species the fingers are in fact jet black (fig. 48), but in some (pl. 45) they are merely dusky grey.

Xanthids range in size from insignificant and hair-obscured species of *Pilumnus* from southern Australia, less than half an inch in width, up to the giant Tasmanian deep-water crab, *Pseudocarcinus gigas* (fig. 47), reaching a width of 16 inches and a weight of 30 pounds. This Tasmanian giant must be considered one of the largest crustaceans known. It appears to be exceeded amongst living brachyurans only by the giant majid spider crab of Japan, *Macrocheira,* which reaches 12 feet across the outstretched arms and 18 inches across the carapace. *Pseudocarcinus* is usually trawled in deep water between about 60 and 100 fathoms on the continental shelf, but has been found almost intertidally in South Australia. It ranges from off central New South Wales through Tasmanian and Victorian waters to south-western Australia. Alive it is uniform deep red in colour or yellow with numerous red patches and irregular markings; the fingers of the asymmetrical hands are black. The larger hand may reach 18 inches from the base of the palm to the tips of the gaping fingers.

The spiny-legged crab, *Actaea peronii* (pl. 43), recalls a beautifully engraved jewel in the delicacy of the distinctive sculpturing on carapace and legs. It occurs in subtidal and shallow waters, usually associated with sponges, off south-eastern and southern Australia from central New South Wales to the Great Australian Bight, South Australia. The attractively-patterned shawl crab, *Atergatis floridus* (fig. 48), is a common and widespread Indo-Pacific species found intertidally in tropical Australia as

Plate 44. Tropical red-eyed crab, *Eriphia sebana,* x1½. *Photo H. Cogger.*

far south on the east coast as Sydney. In the Barrier Reef area it is often seen moving slowly in shallow sandy pools exposed at low tide. The pale lace-like pattern on a smooth, dark green carapace readily identifies this characteristic xanthid crab.

An aggressive and nocturnally active, red-eyed crab, *Eriphia sebana* (pl. 44), is a common xanthid on the Barrier Reef where it favours patches of dead coral and intertidal beach rock. It ranges from East Africa and the Red Sea to Japan, Hawaii and Oceania. In Australia this crab is known from north-western, northern and eastern Australia as far south as central New South Wales. *Eriphia* is one of the few crabs with distinctively coloured eyes. The reticulated coral crab, *Trapezia areolata* (pl. 45), is an unusually-patterned xanthid which lives commensally, and usually in pairs, among the branches of *Pocillopora*-type corals. The reticulated markings on the carapace and chelipeds of this crab resemble the regularly-spaced pattern of polyps on the finger-like branches of these living corals. There are several similar but differently coloured species of *Trapezia* in the Barrier Reef area, each associated with coral growths of approximately the same colour. *Trapezia areolata* occurs with suitable corals from the western Indian Ocean to northern and north-eastern Australia (south to southern Queensland), Japan and Tahiti.

There are extremely few crustaceans which are dangerous when eaten fresh and none which are venomous when handled by human beings. Information on genuinely poisonous crustaceans is very difficult to disentangle from superstition, traditional fear of unusual-looking forms, allergy to crustacean proteins (giving violent reactions in some people) and the effect of toxic food (usually plants) eaten by the crustacean and causing reaction when ingested by a human being. However, there is a recent, well documented case where two children in Fiji died after eating a large specimen of the brown and yellow mottled *Zosimus aeneus* (fig. 49). This is a common and widespread xanthid crab in the Indo-Pacific area and occurs in northern and eastern Australia south to northern New South Wales. It is found under boulders intertidally on the Barrier Reef and has been taken in crayfish pots south of the Reef system. In the present state of our knowledge this crab should not be eaten under any circumstances.

Fig. 49. A large, mottled brown and yellow crab, *Zosimus anaeus*, from tropical Indo-Pacific area, x½. One of the few truly poisonous crabs.

Plate 45. Reticulated coral crab, *Trapezia areolata,* a tropical commensal, x3. 

Freshwater crabs
(Family *Potamidae*)

THE POTAMIDS ARE a relatively small family of exclusively freshwater crabs found in southern Europe, Africa, Asia, the Indo-Malaysian Archipelago, New Guinea, Australia, and in Central and South America. The branchial regions of the carapace are swollen and the body is usually enlarged and transversely oval. Though often referred to as 'river crabs', potamids in general are amphibious in their habits and can be found in rivers, freshwater swamps and burrowing in damp forests.

Six species of the Indian and South-east Asian genus *Parathelphusa* occur in Australia. Five of these are restricted to the Cape York Peninsula, while one, *Parathelphusa transversa* (pl. 46), ranges from inland north-western Australia, across the Northern Territory to the Lake Eyre basin, the Darling River drainage system in western New South Wales and into inland and coastal Queensland (fig. 50). The burrows of this potamid crab

penetrate two to three feet into the clay soil of creek banks, swamps, ground tanks and waterholes. During dry periods, *Parathelphusa* seals the entrance to its burrow with a moulded plug of clay and survives inside till the next wet season. It is difficult to find burrows in habitats which have been dry for any length of time, as weathering of the soil surface destroys earthworks and other signs of excavation. When long periods of drought are broken by heavy rain, observers in inland Australia are often surprised to see active, adult freshwater crabs, or 'sidewalkers' as they are sometimes called, moving about on the surface in large numbers.

As with the sponge crab *Petalomera* (page 80), potamids hatch from large eggs as young crabs and are carried for some time under the abdomen of the mother. The lack of a planktonic larval stage in this 'direct development', means that dispersal and distribution in these crabs depends entirely on the movement of adults. The close relationship of the Australian species, together with their distribution pattern strongly suggests that they entered Australia relatively recently, geologically speaking, from New Guinea. There are many different potamids in that island, but all are derived from ancestral, South-east Asian forms which made their way to New Guinea through the Indonesian chain of islands. During periods of lowered sea level, corresponding to the Pleistocene ice ages, Australia and New Guinea were connected across Torres Strait. The ancestors of *Parathelphusa transversa*, followed by those of the other Australian potamid crabs, crossed to Cape York during these periods and dispersed through the rivers of northern Australia.

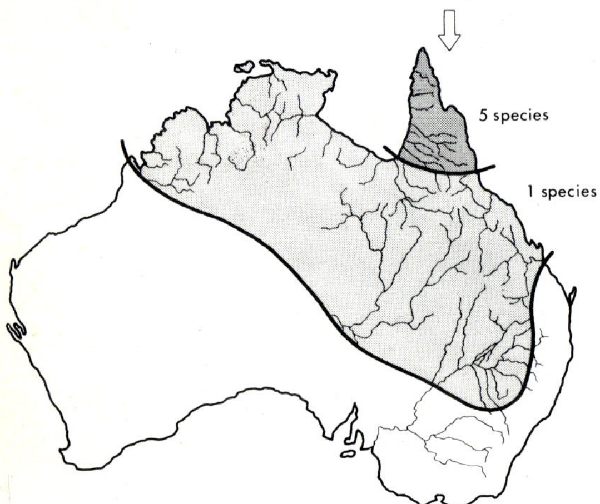

Fig. 50. Distribution of potamid freshwater crabs in Australia. Five species restricted to Cape York Peninsula and one, *Parathelphusa transversa*, widespread in northern and inland eastern river systems.

Plate 46. The inland Australian freshwater crab, *Parathelphusa transversa*, x2.

Shore crabs (Family *Grapsidae*)

T HE COMMON NAME 'shore crab' is often applied to members of both the Grapsidae and the Ocypodidae (page 102). Species of these two families are found all over the world in tropical and temperate regions, under rocks and debris on beaches, on reefs, in estuaries and estuarine marshes, or in burrows along or just above the shoreline. The name 'shore crab' will be used here for the grapsids, or broad-fronted square-backed crabs only, while the ocypodids, both square and oblong-backed, will be referred to as 'long-eyed crabs'.

One of the best known and most highly sought after shore crabs of southern Australia is the red bait crab, *Plagusia chabrus* (fig. 52), which ranges from central New South Wales, south and west to a little north of Perth, Western Australia. It is found near low tide level on exposed rocky coasts, often

Fig. 52. The red bait crab, *Plagusia cnabrus*, from temperate southern Australia, x⅖.

Fig. 51. The 'sourie', *Plagusia glabra*, an eastern Australian intertidal shore crab, x⅔.

submerged and sheltering in seaweed patches, under rock ledges or in crevices. Unmistakable features of the bait crab are the deeply-notched front and the short, dense clothing of hair on the body and legs. It is a dark red crab and may reach four inches across the body. Rock fishermen hunt for *Plagusia* by hand or with thin, barbed spears; it is the favoured bait for reef-dwelling blue grouper in south-Eastern Australia. This large grapsid is also found in South Africa, New Zealand and Chile.

Another species of *Plagusia*, the 'sourie' or *Plagusia glabra* (fig. 51), occurs on eastern Australian shores, but is not used by fishermen as bait. It is smaller than the red bait crab, and the body is smooth and glossy with a characteristic peppering of dark flecks on a fawn and grey background. The swift-footed rock crab or steelback, *Leptograpsus variegatus,* is another common grapsid and probably the most often seen and the most agile rocky shore crab in southern Australia and New Zealand.

Plate 47. Red-fingered marsh crab, *Sesarma erythrodactyla*, green phase, x2.

It varies in colour from a uniform dark olive-green to purple with reddish-yellow central carapace markings.

Marsh crabs of the genus *Sesarma* occur widely in mangrove swamps, river banks and salt marshes throughout the Indo-Pacific area including tropical Australia. They are characterised by a very distinctive, reticulated, grating-like area, on each side wall of the carapace, adjacent to the mouth frame. These areas are made up of intersecting lines of fine hairs and assist in the reoxygenation of the water held in the gill chambers when the crab is on land. The red-fingered marsh crab, *Sesarma erythrodactyla* (pl. 47), is an abundant and conspicuous burrowing crab in eastern Australian mangrove swamps from central Queensland to southern New South Wales. The bright orange-red fingers of the chelipeds and the square carapace distinguish this species from the extremely abundant and rather similar semaphore crab (pl. 50) of the same swamps and mudflats. The red-fingered *Sesarma* is usually greenish-black to nearly black in body colour, but a brilliant, almost iridescent green or greenish-blue colour phase can occur and is shown here on plate 47.

Another common, eastern Australian, estuarine shore crab is the pale honey-brown, almost translucent *Helograpsus haswellianus* (pl. 48). It can be found in sheltered bays and estuaries, and it penetrates well up rivers and into salt marshes surrounding coastal lakes. It lives under rocks or in burrows amongst grass above high tide level, both along river banks and behind mangrove flats. *Helograpsus* ranges from southern Queensland, through New South Wales, Victoria and Tasmania to Spencer Gulf, South Australia. Plate 48 shows this shore crab among dead *Zostera* (eel-grass) leaves above high tide mark on a sandy mudflat in a sheltered, central New South Wales estuary. The sequence of crabs found living on this mudflat between the rocks above the intertidal zone on the left and the soft mud below low tide level on the right is illustrated semi-diagrammatically in figure 53. From the left there is *Helograpsus* among supra-littoral *Zostera* fronds, and then the similar, but smaller, smooth shore crab, *Cyclograpsus audouini,* under stones at high tide mark. In the centre there are displaying males of the equal-handed semaphore crab (pl. 50) and the unequal-handed fiddler crab (front jacket plate). On the right there is a lone soldier crab (pl. 52) and then a sentinel crab (pl. 51) with eye-stalks erect.

Fig. 53. Crab sequence across an intertidal estuarine sandy mudflat in central NSW: shore crabs *Helograpsus* and *Cyclograpsus* among high tidal stones and debris, semaphore and fiddler crabs burrowing in midtidal sand, soldier and sentinel crabs in low tidal mud.

Plate 48. A temperate, estuarine shore crab, *Helograpsus haswellianus,* x3.

Long-eyed crabs
(Family *Ocypodidae*)

OCYPODID CRABS HAVE long eyestalks, often almost half the width of the carapace, and what are described as 'narrow fronts', meaning their eyestalks arise close together at the front of the carapace. The orbits, or grooves in which the eyestalks rest when lying flat, occupy the whole of the anterior margin of the carapace exclusive of the narrow front. These crabs in general live on sandy beaches, mudflats and in mangrove swamps; they are active burrowers and some are social species living in what can be loosely called colonies.

Species of the genus *Ocypode* are called ghost crabs, partly because of their pale colouring and partly from their habit of running swiftly ahead of an observer when disturbed on a sandy beach at night. The most distinctive of the Australian ghost crabs is *Ocypode ceratophthalma* (pl. 49 and fig. 54) with dark, horn-like extensions to the eyestalks in adult specimens. This ocypodid excavates deep burrows on tropical and temperate beaches and, though primarily nocturnal, may at times be found at the surface carrying sand out of a burrow during the day. This horn-eyed ghost crab is a widespread and common species in the tropical Indo-Pacific area and ranges from East Africa and the Red Sea to Japan, Hawaii and Tahiti. In Australia it has been recorded from Shark Bay in Western Australia, through northern and eastern Australia to Shellharbour, southern New South Wales. At least three other ghost crabs occur in Australia and one of these, *Ocypode cordimana*, a small species with

Fig. 54. Horn-eyed ghost crab, *Ocypode ceratophthalma*, a nocturnal, swift-running inhabitant of Australian tropical and warm temperate sandy beaches, x1.

Plate 49. The horn-eyed ghost crab, *Ocypode ceratophthalma*, x2½.

much the same geographic range as the horn-eyed *Ocypode,* can be found burrowing on sandy New South Wales headlands a hundred feet or more above the tide mark and hundreds of yards inland. Occupied burrows have characteristic crab-tracks leading from them to the beach as these ghost crabs forage widely and return to the sea at night.

Another sandy beach ocypodid is the sand-bubbler crab, *Scopimera inflata,* which leaves star-like aggregations of tiny round sand pellets radiating from its burrow at low tide. Estuarine mudflats and mangrove swamps are the habitat par excellence of the social and colonial long-eyed crabs. Figure 55 shows part of an extensive New South Wales estuary system. Semaphore crabs occur under the mangroves, sentinel crabs out on the *Zostera* flat and fiddler crabs are found on the nearby mudflat shown in figure 53.

The semaphore crab, *Heloecius cordiformis* (pl. 50), occurs in great numbers on the eastern and south-eastern Australian coastline between southern Queensland and Tasmania. It is an equal-handed ocypodid with a convex, straight-sided, almost barrel-shaped body. In adults both hands are light purple with pale fingers, but in young *Heloecius* the hands are light orange-red. Semaphore crabs excavate deep vertical burrows, and males defend these at low tide from the aggressive approaches of other males. Fighting in male *Heloecius* is both stylised and ritualistic. The chelipeds are extended forwards in both crabs and the body is raised high off the ground. The combatants then move towards

Fig. 55. A south-eastern Australian mangrove swamp and *Zostera* (eel-grass) mudflat, typical habitat of semaphore, marsh, soldier and sentinel crabs. Ghost nipper burrows are numerous in foreground.

Plate 50. Semaphore crab, *Heloecius cordiformis,* a temperate, estuarine burrower, x3.

Fig. 56. Female fiddler crab, *Uca vocans*, a widely distributed Australian and Indo-Pacific species, x2. Note equally-sized chelae in female, contrasting dramatically with those of male shown on front jacket of book.

one another interlocking their hands and making feeble attempts to grasp opposing legs or bodies. The two crabs then push backwards and forwards until one sinks to the ground and scuttles off. From October to February male *Heloecius* can be seen making characteristic courtship displays near the mouths of their burrows. The body is elevated on outstretched walking legs, the chelipeds are stretched out and raised up above the body as the crab tilts backwards (fig. 53, third crab from left). The chelipeds are then jerked down quickly several times in succession. This 'waving' behaviour is the origin of the name 'semaphore crab'.

Fiddler crabs are characteristic mudflat inhabitants throughout the tropical regions of the world. Males always have one cheliped greatly enlarged and brightly coloured and these are waved, or 'fiddled',

during courtship displays in certain fixed patterns characteristic of each species. The most widespread of the several fiddler crabs in the Australian area is *Uca vocans,* whose beautifully coloured males (front jacket plate) contrast sharply with the drab and unadorned females (fig. 56). This fiddler ranges widely in the Indo-Pacific area and occurs in Western, northern and eastern Australia as far south as Sydney. The sentinel crab, *Macrophthalmus crassipes* (pl. 51), is a distinctively shaped ocypodid common on north-western, northern and eastern Australian mudflats, and is often seen partly buried in the mud of a shallow pool with only the elongate eyestalks breaking the surface of the water. The two males shown among *Zostera* in plate 51 are locked together and pushing against one another in a ritualised fight.

Plate 51. Male sentinel crabs, *Macrophthalmus crassipes,* locked in ritual fight, x2.

Soldier crabs (Family *Mictyridae*)

PERHAPS THE MOST unusual, and certainly one of the most spectacular, of all the Australian crabs is the gregarious soldier crab, *Mictyris longicarpus* (pl. 52). This widespread temperate and tropical species, and two close Australian relatives, are well known in popular literature for their habit of forming large aggregations, or 'armies', of individuals on sandy tidal flats during low tide. A predictable sequence of events precedes the formation of armies. As the tide recedes, the smooth surface of the flat becomes broken by eruptions of tumbled sand, occasional individuals emerge, then suddenly large numbers of crabs appear on the previously empty scene. Unlike the sideways movement of other crabs, the soldiers begin to walk forwards towards the water, feeding on surface organic fragments, and leaving behind pear-shaped sand pellets as they move. Gradually they form into small groups (fig. 57) and, as feeding diminishes, these groups merge into armies containing, on occasion, thousands of individuals. Such aggregations may commonly wander distances as great as 500 yards. As the tide returns, the soldier crabs move towards the upper levels of the flats, break formation and begin to go down into the sand again. Their rapid and distinctive burrowing could be described as corkscrewing their way into the sand, as they dig down with the legs on one side while walking backwards with the legs of the other.

Fig. 57. A detached unit of soldier crabs, *Mictyris longicarpus* negotiates a tidal mudflat in south-eastern Australia.

Plate 52. An individual soldier crab, *Mictyris longicarpus*, x2½.

Location of Photographic Subjects (Plates)

Film key: K = Kodachrome II, KA = Kodachrome II type 'A', EP = Ektachrome Prof., EX = Ektachrome X, EH = High Speed Ektachrome.

Front jacket. Male fiddler crab, *Uca vocans*, Church Point, Pittwater, NSW, x4. (K)

Back jacket. Hinge-beak prawn, *Rhynchocinetes rugulosus*, Long Reef, Collaroy, near Sydney, NSW, x15. (K)

Frontispiece. Careel Bay, Pittwater, NSW. (K)

Plate 1 (page 5). A coral reef prawn, *Saron marmoratus*, Gillett Cay, Swain Reefs, Qld, x1. (K)

Plate 2 (page 7). Eyes and front of ornate marine crayfish, *Panulirus ornatus*, Lake Macquarie, NSW, x3. (K)

Plate 3. E of Mt Ptilotus, Tanami Desert Sanctuary, NT, April, 1965. (K)

Plates 4 & 5. Cootamundra, NSW. Rheinberg lighting. (KA)

Plates 6, 7, 8, 9 & 10. Long Reef, Collaroy, NSW. (K)

Plate 11. Gillett Cay, Qld. Fish is *Plesiops nigracans*. (K)

Plate 12. Long Reef, NSW. (K)

Plate 13. *(Top* and *bottom)* 'Ship Rock', Burraneer Bay, Port Hacking, NSW. (K)

Plate 14. Off Potter Point, Cronulla, NSW, from depth of 80ft. (K)

Plate 15. Shop display, Sydney Fishmarkets, NSW. (K)

Plate 16. Hunter River, Newcastle, NSW. (K)

Plates 17 & 18. *(top* and *bottom)*. Long Reef, NSW. (K)

Plate 19. Long Reef, NSW. *Top left* and *right* (EH), *bottom left* and *right* (K)

Plate 20. Minnie Waters, near Grafton, NSW. (K)

Plate 21. Heron Island, Capricorn Group, Qld. (K)

Plate 22. Cook Strait, NZ, bathypelagic. (K)

Plate 23. *(Left* and *right)* Heron Island, Qld. (K)

Plate 24. Congwong Bay, Botany Bay, NSW, from depth of 18ft. (K)

Plate 25. Heron Island, Qld. (EP)

Plate 26. Heron Island, Qld, reef edge, from depth of 35ft. (EX)

Plate 27. Sydney Harbour, NSW. (K)

Plate 28. Kirrawee, near Royal National Park, NSW. (K)

Plate 29. Terrey Hills, near Sydney, NSW. (K)

Plate 30. Roseville Bridge, Middle Harbour, Sydney, NSW. (K)

Plate 31. Off Gillett Cay, Qld, from depth of 28fms. (K)

Plate 32. Heron Island, Qld. (EP)

Plates 33 & 34. Long Reef, NSW. (K)

Plate 35. *(Top)* Chesterfield Reefs, Coral Sea. *(Bottom)* Gillett Cay, Qld. (K)

Plate 36. Pittwater, NSW, from shallow water. (K)

Plate 37. Off Ballina, NSW. (K)

Plate 38. Long Reef, NSW. (K)

Plate 39. Near Broome, WA, from depth of 120ft. (K)

Plate 40. Long Reef, NSW. (K)

Plate 41. *(Top)* Nobby's Beach, Newcastle, NSW. *(Bottom)* detail of spots and barnacles *Chelonobia patula (left)*; egg mass under abdomen *(right)*. (K)

Plate 42. Long Reef, NSW, with algae *Caulerpa flagelliformis*. (K)

Plate 43. Off Bare Island, La Perouse, NSW, from depth of 35ft. (K)

Plates 44 & 45. Gillett Cay, Qld. (K)

Plate 46. Weilmoringle Station, near Bourke, NSW. (K)

Plate 47. Careel Bay, Pittwater, NSW. (K)

Plate 48. Church Point, Pittwater, NSW. (K)

Plate 49. South West Rocks, Trial Bay, NSW. (K)

Plates 50, 51 & 52. Careel Bay, Pittwater, NSW. (K)

Location of Photographic Subjects (Figures)

All black and white photographs taken on Kodak Panatomic X film unless otherwise stated.

Fig. 1. Inland Australia.

Fig. 3. Water supply dam, Sydney area, NSW.

Fig. 4. Ballina, NSW. (K)

Figs. 5, 7 & 8. Long Reef, Collaroy, near Sydney, NSW.

Fig. 9. Church Point, Pittwater, NSW.

Fig. 10. From fish, Sydney Harbour, NSW.

Fig. 12. 'Ship Rock', Burraneer Bay, Port Hacking, NSW.

Fig. 13. Off Potter Point, Cronulla, NSW, from depth of 80ft.

Fig. 14. Off Gillett Cay, Swain Reefs, Qld, from depth of 38fms. (K)

Fig. 15. On prawn trawler off Ballina, NSW.

Fig. 19. Port Hacking estuary, NSW.

Figs. 20 & 21. Long Reef, NSW.

Fig. 22. Heron Island, Capricorn Group, Qld. *Drawn by B. Bertram from photo by E. Slater.*

Fig. 24. Heron Island, Qld. (K)

Fig. 25. Congwong Bay, Botany Bay, NSW, from depth of 18ft. (K)

Fig. 26. Price Cay, Swain Reefs, Qld. (K)

Fig. 27. Newcastle, NSW.

Fig. 28. Lord Howe Island.

Fig. 29. Off Ballina, NSW.

Fig. 30. Murrumbidgee River, NSW.

Fig. 31. Molonglo River, near Queanbeyan, NSW.

Fig. 32. Careel Bay, Pittwater, NSW.

Fig. 33. Off Flinders Island, Bass Strait.

Fig. 34. Southern Australia.

Fig. 35. Long Reef, NSW.

Fig. 36. Sydney area, NSW.

Fig. 37. Macquarie Island, Southern Ocean, south of Tasmania.

Fig. 38. Off Newcastle, NSW, from depth of about 40fms.

Figs. 39 & 40. Lady Elliott Island, Qld.

Fig. 41. *(Top)* off Bustard Head, Qld. *(Bottom)* Lady Elliott Island, Qld.

Fig. 42. *Redrawn after W. S. Dembowska.*

Fig. 43. Off Tin Can Bay, Qld, from depth of 25fms.

Fig. 44. Port Phillip, Vic.

Fig. 45. One Tree Island, Capricorn Group, Qld. (K)

Fig. 47. Southern Australia. Australian Museum display.

Fig. 48. Gillett Cay, Qld. (K)

Fig. 49. One Tree Island, Qld.

Figs. 51 & 52. Bare Island, La Perouse, NSW.

Fig. 53. Church Point, Pittwater, NSW.

Fig. 54. South West Rocks, Trial Bay, NSW.

Fig. 55. Careel Bay, Pittwater, NSW.

Fig. 56. Church Point, Pittwater, NSW.

Fig. 57. Careel Bay, Pittwater, NSW.

INDEX

Bold figures indicate colour plates or black and white figures

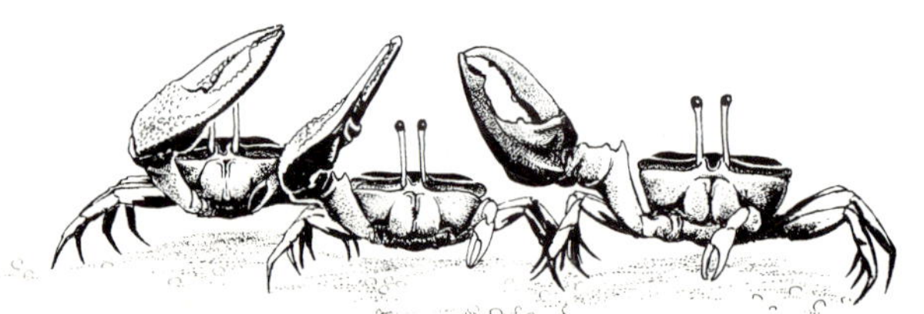